An Introduction to HT
for Scientists and Engineers

David R. Brooks

An Introduction
to HTML and JavaScript
for Scientists and Engineers

 Springer

David R. Brooks, PhD
Institute for Earth Science Research and Education
2686 Overhill Drive
Norristown 19403
brooksdr@drexel.edu

British Library Cataloguing in Publication Data
A catalogue record for this book is available from the British Library

Library of Congress Control Number: 2007926247

ISBN-13: 978-1-84628-656-8   e-ISBN-13: 978-1-84628-657-5

Printed on acid-free paper.

9 8 7 6 5 4 3 2 1

Springer Science+Business Media
springer.com

Preface

i. What is the purpose of this book?

There are many students, other than those specifically interested in computer-related areas such as computer science or computer engineering, who nevertheless need to know how to solve computational problems on computers. There are basically two approaches to meeting the needs of such students. One is to rely on software applications such as spreadsheets, using built-in functions as needed, without relying explicitly on any of the constructs, such as branching and looping, that are common to programming languages.

A second approach is to teach such students a traditional programming language, previously Fortran or Pascal, and more recently C, C++, or Java. These languages play important roles in certain kinds of work, such as computer science (C++, Java) or scientific computing (C and Fortran), but having to learn one of them may be viewed, possibly with good reason, as irrelevant by many students.

From a student's point of view, there is no painless solution to this dilemma, but in this book I assume that learning to solve computational problems in an HTML/JavaScript environment will at least appear to be a more relevant solution. Both **HTML**[1] and **JavaScript** are essential for Web development and some working knowledge of them is a useful and marketable skill. So, in addition to learning basic programming concepts, students can also learn something that may be more immediately valuable than older text-based languages.

In many ways, the HTML/JavaScript environment is more difficult to learn than a traditional text-based programming language such as C. C is a mature (some might prefer "obsolete"), fairly small language with an unambiguous set of syntax rules and a primitive text-based input/output interface. You can view the limitations of C as either a blessing or a curse, depending on your needs. A major advantage of C is that programs written in **ANSI** Standard C should work equally well on any computer that has a C compiler, making the language inherently **platform-independent**.

---

[1] See Glossary for definitions of terms appearing in bold font.

JavaScript and HTML, on the other hand, are immature and very unstable languages (if we can agree informally to call HTML a "language") that function within a constantly changing Web environment. There are dialects of HTML and JavaScript that will work only on particular computing platforms and with specific software. While it is true that there are extensions to languages such as C and other older languages that are platform-dependent, the platform dependence of HTML and JavaScript is a major implementation issue rather than an occasional minor inconvenience.

As one indication of the teaching and learning challenges these environments provide, just three popular paperback HTML and JavaScript reference books occupy nearly 6 inches of space (15 cm in deference to non-U.S. readers) on my office bookshelf! A great deal of the material in those books is devoted to explaining the often subtle differences among various versions of HTML and JavaScript.

Fortunately, it is possible to present some core subsets of both HTML and JavaScript that can be used to solve some of the same kinds of computational problems that would be appropriate for a more traditional language such as C or C++. My motivation for writing this book was to learn how to use HTML and JavaScript to write my own online applications, and I now use this environment for many tasks that I previously would have undertaken in C. Based on this experience, I have concluded that, despite the fact that JavaScript is definitely not intended as a "scientific computing" language, it is nonetheless reasonable to present some basic programming skills of interest to science and engineering students and practitioners in the context of an HTML/JavaScript environment. The examples and exercises presented in the book do not require extensive science, engineering, or mathematics background (only rarely, in a few of the exercises) is knowledge beyond basic algebra needed), so I believe this book could serve as a beginning programming text even for high school students.

ii. Learning by Example

It is well known that people learn new skills in different ways. Personally, I learn best by having a specific goal and then studying examples that seem related to that goal. Once I understand those examples, I can incorporate them into my own work. I have used that learning model in this book, which contains many complete examples that can serve as starting points for your work.

This model works well in an online environment, too. The amount of online information about both HTML and JavaScript is so vast that it is only a slight exaggeration to state that nobody writes original JavaScript code any more. If you have trouble "learning by example," you will have trouble learning these languages, not just from this book, but in general because that is how most of the available information is presented.

It is an inescapable fact that a great deal of the source code behind Web pages involves nothing more (or less) than creative cutting, pasting, and tweaking of existing code. Aside from the issues of plagiarism and intellectual dishonesty that must be dealt with in an academic environment, there is also the practical matter of an effective learning strategy. You cannot learn to solve your own computational problems just by trying to paste together someone else's work. (Believe me, I've tried!) Until you develop your own independent skills, you will be constantly frustrated because you will never find *exactly* what you need to copy and you will be unable to synthesize what you need from what is available.

So, while you should expect to find yourself constantly recycling your own code throughout a course based on this book, you need to be responsible for your own work. Make sure you really *learn* and don't just *learn to copy*!

iii. Acknowledgments

I am once again indebted to my wife, Susan, for her patient editing of this manuscript. Considering that she also edited two of my previous computer programming manuscripts, I can conclude only that sufficient time has passed to dim her recollections of those experiences!

During the Fall of 2006, student comments in a class I taught for Drexel University's School of Biomedical Engineering, Science and Health Systems, have provided valuable suggestions for improving the presentation and content of this manuscript.

Contents

1. Introductory Concepts

Chapter 1 provides a very brief introduction to using HTML and JavaScript for creating simple Web pages. It presents examples that illustrate the way in which JavaScript interfaces with an HTML document to display some printed output in a Web browser window, and introduces the concept of an HTML document as an object, with certain methods and properties accessible through JavaScript to act on that object. There are also some examples that show how to modify the appearance of a document by using HTML tags and their attributes, including as part of a text string passed as a calling argument to JavaScript's `write()` method.

1.1 Introducing the Tools

1.1.1 What Is an HTML Document?

HTML is an acronym for **H**yper**T**ext **M**arkup **L**anguage. **HTML documents**, the foundation of all content appearing on the **World Wide Web (WWW)**, consist of two essential parts: information content and a set of instructions that tells a computer how to display that content. The instructions—the "markup," in editorial jargon—comprise the HTML language. It is not a programming language in the traditional sense, but rather a set of instructions about how to display content. The computer application that translates this description is called a **Web browser**. Ideally, online content should always look the same regardless of the browser used or the operating system on which it resides, but the goal of platform independence is achieved only approximately in practice.

A basic HTML document requires a minimum of four sets of **elements**:

```
<html> ... </html>
<head> ... </head>
<title> ... </title>
<body> ... </body>
```

These elements define the essential parts of an HTML document: the document itself, a heading section, a title section, and a body. Each of the elements is defined by two **tags**—a start tag and an end tag. Tags are always enclosed in angle brackets: <...>. End tags start with a slash (/). As is shown later, some HTML elements have only one tag. Most tags are *supposed* to occur in pairs, although this rule is only loosely enforced in HTML. In order to support a **scripting language** such as JavaScript (much more about that later!), another element must be added:

```
<script> … </script>
```

For our purposes, a `script` element always contains JavaScript code.

These elements are organized as follows within an HTML document:

```
<html>
  <head>
    <title> … </title>
    …
    <!-- Optional script elements as needed. -->
    <script> … </script>
  </head>
  <body>
    …
  </body>
</html>
```

The `html` tag encloses all other tags and defines the boundaries of the HTML document. We will return to all the other tags later. `script` tags are often found inside the <head> tag, but they can appear elsewhere in a document as well. The indenting used to set off pairs of tags is optional, but it makes documents easier to create, read, and edit. This style is part of good programming practice in all languages.

Owing to the fact that JavaScript is so tightly bound to HTML documents, you must learn JavaScript along with at least a subset of HTML. Unfortunately for anyone trying to learn and use HTML and JavaScript, each of the several available browsers is free to implement and support JavaScript in its own way. A browser does not even have to support JavaScript at all, although it is hard to imagine why it would not do so. Browsers can and do incorporate some proprietary HTML and JavaScript features that may not be supported by other browsers. Newer versions of any browser may support features that will not be recognized by earlier versions.

Fortunately, it is possible to work with what is essentially a de facto standardized subset of HTML and JavaScript. As a result, some of the descriptions of the details of HTML and JavaScript in this book are incomplete. This is not necessarily bad!

Although we tend to think of HTML documents as a way to distribute information for remote access on the Web, they are equally useful when used locally on any computer that has a browser. Thus, in conjunction with JavaScript, you can create a self-contained problem-solving environment that can be used locally as well as (literally) globally.

Good programming technique often involves separating the **input/output (I/O) interface** from the underlying calculations that do the work of a program, using appropriate modularization. The programming environment provided by HTML/JavaScript provides a conceptually elegant means of implementing this strategy. An HTML document provides the I/O interface and JavaScript handles the calculations. An advantage of HTML is that it provides a wealth of interface possibilities that far surpass those of text-based languages such as C.

1.1.2 What Is JavaScript?

JavaScript is an **interpreted** (rather than a **compiled**) **object-oriented programming language**, with roots in C/C++, that has been developed for use with other Web tools. It does not operate as a standalone language, but rather is designed to work together with HTML for creating interactive Web pages. JavaScript is not the same as Java, which is a compiled object-oriented language.

JavaScript is used to write **client side applications**, which means that its code is sent to a user's computer when a Web page is loaded. The code is then executed, basically line by line, by a JavaScript interpreter included as part of the user's (client's) Web browser. This arrangement minimizes security issues that can arise when a client computer interacts with the computer that sent the page. It also makes it easy to package an entire problem—with its own user interface and solution—self-contained within a single document. However, the inability to interact dynamically with information on the server does impose limitations on the kinds of tasks that JavaScript can accomplish.

It is commonplace to refer to any set of written computer instructions as a "program," but this term should perhaps be more rigorously applied to a separate entity that can be executed on its own. As JavaScript is interpreted rather than compiled, a separately executable entity is never created. Rather, JavaScript code statements are interpreted and executed one at a time, essentially "on the fly." Although this may seem inefficient, there is rarely any discernible time lag associated with executing JavaScript commands on modern computers.

JavaScript is one of a class of scripting languages whose purpose is to access and modify components of an existing information interface. (Microsoft's VBScript is another scripting language.) In this case, the interface is an HTML document. Something like JavaScript became necessary as soon as HTML documents on the Web evolved from one-way delivery systems for displaying fixed content. One of JavaScript's first applications arose from the need to check values entered by users into the fields of HTML forms that can be sent back to the originator. (Forms are discussed in a later chapter.) JavaScript can be used to compare input values against an expected range or set of values and to generate appropriate messages and other actions based on those comparisons.

JavaScript has evolved into a complete programming language with extensive capabilities for manipulating text and handling mathematical operations, useful for a wide range of computing problems. The possible applications include many self-contained scientific and engineering calculations, which provide the primary motivation for this book. As noted above, the utility of JavaScript is restricted to problems that do not have to access external data sources, such as would reside on a host computer and would not be available to a client computer.

The major challenge in learning HTML/JavaScript is that it is not a completely standardized environment. The various dialects of HTML and JavaScript pose problems even for experienced programmers. These kinds of problems can be minimized by focusing on an appropriate subset of HTML/JavaScript, which is feasible because there is little reason to use browser-specific subsets of HTML/JavaScript in the context of the topics dealt with in this book.

1.1.3 How Do You Create HTML/JavaScript Documents?

Since HTML/JavaScript documents are just text documents, they can be created with any text editor. Even Windows' very basic Notepad application is a workable choice for simple tasks.[1] Once they are created, you can open HTML files in your computer's browser—hopefully without regard to which browser you are using. As long as you give such documents an `.htm` or `.html` file name extension, they should automatically open in

---

[1] When you save a file in Notepad, the default extension is `.txt`. You may have to enclose the file name with an `.htm` extension in quote marks to prevent Notepad from adding the `.txt` extension.

your browser when you double-click on the file name. The three-letter extension is standard for Windows-based documents. The four-letter extension is commonly used on UNIX systems.[2]

There is one other consequence of using Windows computers for creating all the examples in this text (and the text itself, for that matter): Windows file names are case-insensitive, whereas in UNIX, all spellings, including file names and commands, are case-sensitive. This should not cause problems, but it is something to keep in mind. In Windows, you can name a document `newDocument.htm`. Later, you can spell it `newdocument.htm`, `NEWDOCUMENT.HTM`, or any other combination of uppercase and lowercase letters and it will not matter. However, on a UNIX system, that file can be accessed only with the original spelling.

Although you can create text (and, therefore, HTML) documents with a full-featured word processor such as Microsoft Word, this is not recommended. When you save a word processor document it no longer contains just the text you have typed, but also all the layout and formatting information that goes with along with it. You can choose to save a document as just text with an `.htm` extension, but it is easy to forget to do it.

Microsoft Word and other modern word-processing applications can also format any document as an HTML document, but this is also not recommended. These converted documents may include a huge quantity of extraneous information and HTML instructions that make the resulting file much larger and more complex than it needs to be. (Try saving a Word document as an HTML document and then look at the result in a text editor such as Notepad!)

RTF ("rich text format") documents are also unacceptable, as they still retain some formatting information that is inappropriate for an HTML document. Any document that contains "smart quotes" rather than "straight quotes" can also cause problems, because smart quotes may not be displayed properly by browsers. (This is much less of a problem on current browsers than it was in the past.)

There are commercial Web development applications that allow you to create Web pages without actually knowing anything about HTML or JavaScript, but these applications are not suitable for use with this book. The obvious reason is that the primary purpose of the book is to

text Editors vs. Word Processors

---

[2] On Windows computers, you can associate extensions with whatever application you wish. So, for example, if you have more than one browser installed on your computer, you can designate one of them as the default browser and assign it as the application for opening HTML documents.

show you how to write your own HTML instructions and JavaScript code. Moreover, these applications will probably create HTML files that are much larger and more complex than they need to be for our purposes. Finally, they do not include the kind of JavaScript code you will need for the topics discussed here. Thus, they are better suited for Web development projects that involve a lot of graphics and the other "bells and whistles" that make commercial Web pages attractive.

Creating an HTML/JavaScript document that works properly inevitably involves switching back and forth between a text editor and a browser—making changes and observing the effects of those changes. Once you create a basic HTML document, you can open it in your browser and move back and forth between this document and your text editor, and whenever you change the document, you can reload or refresh it in your browser. It is certainly possible, but not particularly convenient, to do this with a simple text editor such as Notepad.

There are many commercial software tools whose purpose is to facilitate writing and editing HTML documents by integrating document creation, editing, and viewing. Some of them are intended for large and complicated projects and may be "overkill" for use with this book. For several years, for creating this book and in my own day-to-day work, I have used AceHTML Freeware V.5 (see www.visicommedia.com). This software provides an HTML/JavaScript editor with some automatic color-based text formatting that makes HTML instructions and JavaScript code easy to read. There is an integrated Web page viewer, so it is easy to switch back and forth between creating and editing a document and seeing the results of your work. It also has a JavaScript syntax checker. As is typically the case, the checker is not very good at telling you how to fix a syntax error, but it at least tells you where the error was detected. The freeware version of this editor may or may not be currently available, and it may require installation of other software that you may or may not want on your computer. In any event, during the time I was writing this book, there were versions of AceHTML available for purchase.[3]

Although, in principle, it *should not* make any difference which browser you use, the outputs I have displayed in this text come from either AceHTML's internal browser or Mozilla's Firefox, which I use as the default browser on my Windows computers. When you display content in

---

[3] Recent versions of AceHTML assume XHTML as the default language, rather than HTML. If you use such a version with this book, you must override this assumption by saving files with .htm or .html extensions.

an "alert" box, as will be described later in this book, the appearance of this box is different for different browsers, and hence may be different from what is shown here.

1.1.4 Some Typographic Conventions Used in This Book

HTML tags and JavaScript code are printed in a monospaced (Courier) font in document examples and whenever they are referred to in the text. Thus, document is interpreted as a reference to an HTML **object**, as opposed to its general use as a term identifying a body of text. Some technical terms used for the first time are printed in **bold font**. Their definitions can be found in the Glossary. Within descriptions of HTML document features and JavaScript code, user-supplied text is denoted by *{italicized text in braces (curly brackets)}*. In the code examples, HTML tags are printed in **bold font.**

The renderings of HTML documents and other output as displayed in a browser window have been captured and edited on a Windows computer by pressing the PrtScn (or Print Screen) key and copying the resulting screen image into the freeware IrfanView image editing program (www.irfanview.com).

Owing to the small format of this book, line breaks in document examples may sometimes be misleading. I have tried to make necessary line breaks as logical as possible, but you may want to remove some breaks when you reproduce these documents for your own use.

1.1.5 Where Should I Look for More Information about HTML and JavaScript?

By now, it should be clear that this book is in no way intended as a reference source for either HTML or JavaScript. Any attempt to provide complete coverage for either language would thoroughly confound its purpose and is far beyond my capabilities! Therefore, you must look elsewhere for exhaustive treatments of HTML and JavaScript. I used three sources:

Thomas Powell, *HTML: The Complete Reference, Third Edition*, 2001, Osborne/McGraw-Hill, Berkeley, CA. ISBN 0-07-212951-4.

Thomas Powell and Dan Whitworth, *HTML Programmer's Reference, Second Edition*, 2001, Osborne/McGraw-Hill, Berkeley, CA. ISBN 0-07-213232-9.

Thomas Powell and Fritz Schneider, *JavaScript: The Complete Reference*, 2001, Osborne/McGraw-Hill, Berkeley, CA. ISBN 0-07-219127-9.

   If you are at all serious about creating your own online applications ("serious" perhaps being defined as anything past the bare minimum needed to survive a course based on this text), there is no substitute for these or similar references.

   The first HTML book I ever read (and still consult from time to time) is out of print, but it is worth looking for in libraries or remaindered book stores (which is where I found my copy). Even though it addresses an older (and simpler!) version of HTML, it is still an excellent resource for the kinds of problems discussed here.

Todd Stauffer, *Using HTML 3.2, Second Edition*, 1996, Que Corporation, Indianapolis, IN. ISBN 0-7897-0985-6.

1.2 Your First HTML/JavaScript Documents

A typical first goal in learning any programming language is to display a simple message. With HTML, this is trivially simple: Just type the message in the body of the document, as shown in Document 1.1. (Appendix 1 contains an index to all the documents in the text.)

Document 1.1 (`HelloWorldHTML.htm`)

```
<html>
<head>
<title>First HTML Document</title>
</head>
<body>
Hello, world!
</body>
</html>
```

```
Hello, world!
```

   Most document examples presented in this book will include a browser's rendering of the screen output produced by the document. When a border appears around the output, as it does for the output from Document 1.1, the purpose is to distinguish the output from the rest of the text—the document does not generate that border. In the text, renderings are always in black and white or grayscale. In some cases, as noted, color renderings are printed on separate color plates. In other cases (such as for Document 1.3) you will have to try the code yourself.

Document 1.1 is certainly not very exciting. But the point is that an HTML document simply displays the static content you provide. As you will learn in Chapter 2, HTML provides many facilities for changing the *appearance* of this content, but not the content itself.

You can also display content with JavaScript. With JavaScript, input and output always pass through an HTML document. Instructions (code) you write in JavaScript are called a **script**. The capability to interpret JavaScript instructions must be built into your browser. Document 1.2 uses JavaScript to generate a simple text message, which is displayed in the document. There is no good reason to use JavaScript simply to display fixed content, but this exercise will provide an introduction to JavaScript syntax. Do not worry if the details of this and following examples seem obscure—hopefully, future chapters will clarify all these details!

Document 1.2 (`HelloWorld.htm`)

```
<html>
<head>
 <title>Hello, world!</title>
 <script language="javascript" type="text/javascript">
 // These statements display text in a document.
   document.write("Hello, world!");
   document.write("<br />It's a beautiful day!");
 </script>
</head>
<body>
<!-- No content in the body of this document. -->
</body>
</html>
```

> Hello, world!
> It's a beautiful day!

A browser must be instructed to interpret certain parts of an HTML document as JavaScript code. To accomplish this, all text appearing inside the `script` element will be interpreted by a browser as one or more JavaScript statements. This means that HTML elements cannot appear inside the `script` element, as then the JavaScript interpreter would attempt (inappropriately) to interpret them as JavaScript code. This will generate a JavaScript error. In Document 1.2, the `<br />` tag, which generates a line break, is an HTML element, but it is included inside a quoted string of text. This is allowed, but

```
document.write("Hello, world!");
<br /> document.write("It's a beautiful day!");
```

is not allowed.

As noted previously, JavaScript is an object-based language. In programming terminology, an HTML document is an object. Using JavaScript, pre-defined **methods** can be used to act on a specified object. (Objects are discussed in more detail starting in Chapter 4.) Document 1.2 accesses ("calls" or "invokes") the `write()` method of the `document` object to display text. A method is associated with its object by using "dot notation," as in `document.write()`.

Methods such as `write()` often, but not always, require one or more inputs, referred to as **calling arguments**. In Document 1.2, the text strings `"Hello, world!"` and `"<br />It's a beautiful day! ";` are calling arguments for the `write()` method. Calling arguments provide the values on which a method acts.

As we will see, most HTML elements include **attributes** that are used to assign properties to the element. The `script` element *should* include values for the `language` and `type` attributes, as shown:

```
<script language="javascript" type="text/javascript">
```

Comments within an HTML document are indicated by a very specific sequence of symbols:

```
<!-- {comments} -->
```

In keeping with the style adopted in this book, italicized text enclosed in curly brackets indicates text that is entered by the user. The curly brackets *could* be part of the comment, but are not needed and would normally not be included.

Inside a `script` element, single-line comments begin with two slashes, as in the fifth line of Document 1.2. Comments are a basic part of good programming style, no matter what the language. Some authors prefer not to use many comments in HTML/JavaScript because it increases the size of the file that is sent to the client computer. However, when you are learning the material presented in this book, there is no excuse for not making liberal use of comments to remind yourself of what you are doing.

One use of HTML comments is to hide JavaScript code from browsers that do not have a JavaScript interpreter, but this is much less of a problem today than it might have been several years ago. It is also irrelevant for now because, of course, your browser must support JavaScript in order to be useful for this book. In any event, hiding JavaScript is accomplished as follows:

```
<script language="javascript" type="text/javascript">
  <!-- Start hiding JavaScript code here.
      {Put JavaScript statements here.}
  // Stop hiding code here. -->
</script>
```

Although these HTML comment tags appear to be out of place because we have already stated that HTML elements cannot appear inside a `script` element, any browser that includes a JavaScript interpreter will be able to sort things out, basically by ignoring the comment tags.

HTML syntax is case-insensitive, which means that `<html>` is equivalent to `<HTML>` or even `<hTmL>`. Some HTML document authors favor uppercase spellings for tags because they stand out from the text content. However, **XHTML** (extended HTML), the apparent successor to HTML, requires that tags be in lowercase letters.[4] Hence, I always use lowercase letters for tag names here. Note that, despite previous warnings that file names and commands are case-sensitive in some systems, browsers should not be case-sensitive in their interpretation of HTML tags, regardless of the underlying operating system.

JavaScript syntax is *always* case-sensitive, regardless of the computer system on which it runs, like the C/C++ languages from which it is derived. So, when you write JavaScript code, you have to be very careful about case. For example, `document` is an object name recognized by JavaScript, but `Document` is not. (Try this in Document 1.2 if you need convincing.)

Note that each of the two JavaScript statements (the calls to `document.write()`) is terminated with a semicolon. JavaScript interprets a semicolon as "end of statement." As a matter of syntax, a line feed at the end of a statement will also be interpreted as marking the end of that statement. However, it is poor programming practice to use this "implied semicolon," and all JavaScript statements used in this book *should* terminate with semicolons. (Authors are not perfect!)

You can make Document 1.2 a little fancier by using other HTML elements and their attributes to control the appearance of the text. (Chapter 2 presents much more information about elements and attributes.) In Document 1.3, `font` (font description), `h1` (heading), and `hr` (horizontal rule) are elements, and `color`, `size`, and `align` are attributes. Of these,

---

[4] Although this book adopts some XHTML style rules, the documents are written in HTML and are not intended to be fully XHTML-compliant.

the hr element requires only a single tag because it does not enclose any HTML content. Single-tag elements should include a forward slash at the end: <hr /> rather than <hr>.

Document 1.3 (HelloWorld2.htm)

```
<html>
<head>
<title>Hello, world!</title>
</head>
<body>
<h1 align="center">First JavaScript</h1>
<hr />
<script language="javascript" type="text/javascript">
   document.write("<font size='5'
     color='red'><center>Hello, world!</font>");
   document.write("<br /><font size='7' color='blue'>
     It's a beautiful day!</center></font>");
</script>
</body>
</html>
```

First JavaScript

Hello, world!
It's a beautiful day!

(Try this yourself to see the colors displayed.)

As previously noted, there is no good reason to use JavaScript to display this fixed content, but Document 1.3 again makes the point that any HTML tags appearing as part of the calling argument passed to document.write() are treated as part of the text string—the characters enclosed in quote marks—and therefore do not violate the rule that HTML elements cannot be used inside a script element. The HTML tags are essentially "pasted" into the HTML document right along with the text. Within the string

```
"<br /><font size='7' color='blue'>
  It's a beautiful day!</center></font>"
```

the attribute values are enclosed in single rather than double quotes. Otherwise, it would not be clear where the quoted string begins and ends.

Another difference between Document 1.2 and Document 1.3 is that in 1.3, the script element is inside the body element. This is all right, although we would normally try to keep the script element inside the head element, thus ensuring that the JavaScript code is interpreted before the rest of the page is loaded. This detail is of no concern in this example, the sole purpose of which is to display some text.

As expected, this attempted modification of the script, which contains HTML tags in a context where a browser expects to see only JavaScript code, will produce an error:

```
<script language="javascript" type="text/javascript">
  <font size="5" color="red"><center> // ERROR!!
  document.write("Hello, world");
  </font>
</script>
```

You can include more than one script element within an HTML document, as shown in Document 1.4a, in which there are two separate script sections, arbitrarily divided into a section above the horizontal rule (see the <hr /> tag) and another below the rule.

Document 1.4a (HelloWorld3.htm)

```
<html>
<head>
<title>Hello, world! (v.3)</title>
</head>
<body bgcolor="lightgreen" text="magenta">
<h1 align="center">First JavaScript</h1>
<script language="javascript" type="text/javascript">
  document.write("<font color='green'>
  This document was last modified on
  "+document.lastModified+"</font>");
</script>
<hr />
<script language="javascript" type="text/javascript">
  document.write("background = "+document.bgColor);
  document.write("<br />font = " + document.fgColor);
  document.write("<font size='5'
    color='red'><center>Hello,world!</font><br />");
  document.write("<font size='7' color='blue'>
    He said, &quot;It's a beautiful day!&quot;
    </center></font>");
</script>
</body>
</html>
```

First JavaScript

This document was last modified on 05/03/2006 13:34:57

background = #90ee90
font = #ff00ff

Hello, world!

He said, "It's a beautiful day!"

(See Color Example 1 for full-color output.)

Document 1.4a contains an answer to this question: How do you display double quote marks with the `document.write()` method if you cannot use double quotes inside a quoted string? One way is to use the **escape sequence** `&quot;`. Escape sequences always start with an ampersand (`&`) and end with a semicolon (`;`). There are many escape sequences for displaying characters that are not available directly from the keyboard or would be misinterpreted by HTML if entered directly, and they will be discussed later as needed. A list of commonly used escape sequences appears in Appendix 2.

JavaScript objects have **properties** as well as methods. Like methods, properties are associated with objects through the use of dot notation. One useful property of the `document` object is `lastModified`, used in Document 1.4a. As its name suggests, this property accesses the time and date stamp automatically stored along with a document whenever it is modified and saved, based on the calendar and clock on the computer used to create the document. This stamp is automatically attached to the document, without any special action required by the creator of the document. The `lastModified` property is useful for documents that contain time-sensitive information, or just to give users some idea of whether a page displayed in a browser is current.

Document 1.4a contains the following two statements, which access two more `document` properties:

```
document.write("background = "+document.bgColor);
document.write("<br />font = " + document.fgColor);
```

These display a code for the background and font colors.

Attributes such as `size` and `color` have values. These values are *supposed* to be enclosed in quotes, although this is not actually required in HTML. Quotes *are* required in XHTML, and we will always use them. You can use either double or single quotes. In HTML documents, double quotes are generally accepted as the standard. However, when HTML elements with attributes are included inside quoted strings, as in

```
document.write("<font size='5'
  color='red'><center>Hello,world!</font><br />");
document.write("<font size='7' color='blue'>
  He said, &quot;It's a beautiful day!&quot;
  </center></font>");
```

then single quotes are required for the values in order to avoid conflict with the double quotes around the string.

A more reasonable approach to generating the output shown for Document 1.4a is to use JavaScript only as required to access desired document properties (and perhaps display some related text) and to use HTML for everything else. Document 1.4b is a modified version of Document 1.4a that does the content formatting with HTML tags inside the document. There is no need to show the output, as it is identical to that for Document 1.4a.

Document 1.4b (HelloWorld3HTML.htm)

```
<html>
<head>
<title>Hello, world! (with HTML)</title>
<script language="javascript" type="text/javascript">
  document.write(
  "<font color='green'> This document was last modified on
  "+document.lastModified+"</font>");
</script>
</head>
<body bgcolor="lightgreen" text="magenta">
<h1 align="center">First JavaScript</h1>
<hr />
<script language="javascript" type="text/javascript">
  document.write("background = "+document.bgColor);
  document.write("<br />font = " + document.fgColor);
</script>
<font size="5" color="red"><center>Hello,world!</font><br />
<font size="7" color="blue">
He said, "It's a beautiful day! "</center></font>"
</body>
</html>
```

In this case, there is actually a justification for putting one of the script sections inside the body of the document: This script is used to display codes for the background and text colors, which are known only after they are set inside the `body` element.

A summary of some properties and methods of the `document` object is given in Table 1.1. The `bgColor` and `fgColor` properties are not used to *set* the colors, but merely to tell you what they are. (You are right to conclude that this is normally not terribly important information.) Note that `bgcolor` is an HTML attribute used to set the background color of the `body` element and is supposed to be (but does not have to be in case-insensitive HTML) spelled in lowercase letters. `bgColor` is a property of the JavaScript `document` object and must be spelled with a capital C, as shown.

Table 1.1. Some properties and methods of the `document` object

Property or Method	Action
Property `document.bgColor`	Returns current value of background (page) color. Returns `"#ffffff"` for `<body bgcolor="white">`
Property `document.fgColor`	Returns current value of font color. Returns `"#0000ff"` for `<body text="blue">`
Property `document.lastModified`	Returns text string containing date the document was last modified.
Method `document.write("Hello!")`	Prints quoted string on document page.
Method `document.writeln("Hello!")`	Prints quoted string on document page, followed by line feed.[*]

[*] As HTML ignores line feeds, the `writeln()` method will not normally produce any noticeable difference in output. If the text to be displayed is within a `pre` element, then the line feed will be displayed.

1.3 Accessing HTML Documents on the Web

Documents intended for access by others on the World Wide Web are posted on a **Web server**, a computer system connected to the **Internet**. Colleges and universities typically provide Web servers for use by their faculty and students. Individuals not affiliated with an institution may have to purchase space on a commercial Web server. In any case, access to Web pages is universal in the sense that any computer with an Internet connection and a browser can be connected to a Web site through its Internet address—its Uniform Resource Locator (**URL**).

Not all HTML documents have to be publicly accessible on the Web. They can be protected with logon identifications and passwords, or they can be available only locally through an **intranet** (as opposed to the Internet). The Internet is a global **network** of interconnected computers, whereas an intranet is a local network that may or may not also provide connections to the Internet. For example, a company can provide an intranet with no external access, exclusively for internal use by its own employees.

Internet addresses look something like this:

```
http://www.myUniversity.edu/~myName/index.htm
```

They start with the `http://` prefix, to indicate that the Hypertext Trans-fer Protocol (**HTTP**) is being used. There are some variations, such as `https`, which indicates that the address that follows resides on a secure server, as required for financial transactions, for example. The rest of the address identifies a Web server and then a folder or directory on a com-puter system at `myUniversity` for someone named `myName`. The `.edu` extension identifies this site as belonging to an educational institution, in the same way as `.gov` and `.com` identify government and commercial sites. The ~ symbol is often used to specify a folder (or directory) set aside for Web pages, but there are many ways to specify the location of Web pages. Sometimes names in URLs are case-sensitive, depending on the operating system installed on the computer system containing the Web page. Thus, if you type `myname` instead of `myName` in the above URL, it may not work. Users of Windows computers should note the use of for-ward slashes rather than backslashes to separate folders (or directories).

The `index.htm` (or `index.html`) file contains the **home page** for this individual. By default, the `index.htm` file is automatically opened, if it exists, whenever this URL is accessed. That is, the address

```
http://www.myUniversity.edu/~myName/
```

is equivalent to the address that includes the `index.htm` file name.

As they were being developed, the documents discussed in this book resided neither on the Internet nor on an intranet. Rather, they were simply stored in a folder on a computer and accessed through the file menu in a browser, just as you would access a file with any other software application. For example, the "address" on my computer for the first document in this text is

```
file:///C:/Documents%20and%20Settings/David/Desktop/
JavaScript/JavaScriptCode/HelloWorld.htm
```

(Spaces are represented by the hexadecimal code %20 and, yes, there are three forward slashes following file:)

You should create a separate folder on your computer as you work through the examples in this book and write your own documents. You *could* make documents you create yourself accessible on the Internet or an intranet by placing them on a Web server. For example, if you are taking a course based on this book, your instructor may require you to post homework assignments on a Web site.

1.4 Another Example

The following example shows how to include an image in an HTML document.

Document 1.5 (house.htm)

```
<html>
<head>
<title>Our New House</title>
<script language="javascript" type="text/javascript">
document.write("<font color='green'>This document was
  last modified on "+document.lastModified+"</font>");
</script>
</head>
<body>
<h1>Our New House</h1>
<p>
Here's the status of our new house. (We know you're
fascinated!)</p>
<!—Link to your image goes here. -->
<img src="house.jpg" align="left" /><br />
</body>
</html>
```

There are several image formats that are widely used in HTML documents, including image bitmaps (.bmp), Graphics Interchange Format (.gif), and Joint Photographic Experts Group (.jpg).

The original .jpg file has been compressed to reduce its size, and this compression can result in jagged edges where edges should be straight. This effect is visible in the house framing and roof lines.

This document was last modified on 05/03/2006 13:12:30

Our New House

Here's the status of our new house. (We know you're fascinated!)

Within the img element, height and width attributes allow you to control the size of the image display (in pixels). This is not equivalent to actually "resizing" the image, as is possible with image-editing software.[5] Hence,

(See Color Example 2 for full-color output.)

it is important to use images that initially are sized appropriately. If a very large high-resolution image file is displayed as a very small image, using the height and width attributes, the original large file must still be transmitted to the client computer. In view of the fact that high-resolution images can produce very large files (>10 Mb), it is important to consider appropriate resolution for images included in HTML documents, even in an age of high-speed broadband Internet connections. (The size of the compressed grayscale house.jpg image printed here is about 93 Kb.)

Document 1.5 could be made into a default home page simply by changing its name to index.htm.

Here is a final admonition that I hope does not sound too preachy: Intellectual honesty and fairness in the use of other people's material is important, no matter what the setting. The image displayed in Document 1.5 was taken by me, of my own house under construction. In other

---

[5] I have used IrfanView (www.irfanview.com) for all the image processing in this book. This very popular freeware program does an excellent job of resizing images while maintaining detail from the original image. Of course, I cannot guarantee its availability to my readers.

words, I "own" this image. Whenever you post images (or other material, for that matter) online, please be careful to respect intellectual property rights. Your default approach should be that online materials are copyrighted and cannot be used freely without permission. If you are in doubt about whether you have permission to use an image or other material, don't!

2. HTML Document Basics

Chapter 2 describes the characteristics of an HTML document, including some of the basic HTML elements and their attributes. The list of attributes is not necessarily complete, but rather includes a subset that is used in this book. The chapter includes a description of how to set colors in documents and a brief introduction to cascading style sheets.

2.1 Documents, Elements, Attributes, and Values

2.1.1 Essential Elements

As noted in Chapter 1, JavaScript needs an HTML document to serve as a user interface. Or, stated the other way around, HTML documents need a scripting language such as JavaScript to manage interactions with users. A basic HTML document consists of four sections defined by four sets of elements, arranged as follows:

```
<html>
      <head>
            <title> ... </title>
            ...
      </head>
      <body>
            ...
      </body>
</html>
```

Each of these elements has a start tag and an end tag. Tags are always enclosed in angle brackets <...> and the end tag always includes a forward slash before the element name. The body element supports attributes that can be used to control the overall appearance of an HTML document. Documents, elements, attributes, and values are organized in a specific hierarchy:

HTML document → elements → attributes → values

Elements exist within a document. Elements can have attributes and attributes (usually) have values. Note that some elements are nested inside others. For example, all the other elements are nested inside the `html` element, and the `title` element is nested inside the `head` element.

Following is a brief description of the four elements that will be part of every HTML document. Attributes, if any, are listed for each element. Note, however, that not all the possible attributes are listed. Thus, a listing of "none" may mean that there are attributes for this element, but that they are not used in this book. Consult an HTML reference manual for a complete list of attributes. As several elements can share common attributes, attributes and their values are listed separately, following the list of elements.

`<html> ... </html>`
The `html` element surrounds the entire document. All other HTML elements are nested within this element.
Attributes: none

`<head> ... </head>`
The `head` element contains information about the document. The `head` element must contain a `title` element and under XHTML rules, the `title` must be the first element after `head`. From our perspective, the other important element to be included in `head` is `script`, which will contain JavaScript code.
Attributes: none

`<title> ... </title>`
The `title` element contains the text that will be displayed in the browser's title bar. Every HTML document should have a title, included as the first element inside the `head` element.
Attributes: none

`<body> ... </body>`
The `body` element contains the HTML document content, along with whatever elements are required to format, access, and manipulate the content.
Attributes: `background, bgcolor, text`

2.1.2 Some Other Important Elements

The four basic elements discussed above constitute no more than a blank template for an HTML document. Other elements are needed to display and control the appearance of content within the document. Following are some important elements that you will use over and over again in your HTML documents, listed in alphabetical order. The list of attributes is not necessarily complete, but includes only those that are used in this book.

<a> ... </a>
    The a (for "anchor") element provides links to an external resource or to an internal link within a document.
Attributes: href, name

<b> ... </b>
    The b element forces the included text to be displayed in a bold font. This is a "physical element" in the sense that it is associated specifically with displaying text in a bold font, even though the actual appearance may depend on the browser and computer used. In contrast, see the strong element below.
Attributes: none

 or
    The br element inserts a break (line feed) in the text. Multiple breaks can be used to insert multiple blank lines between sections of text. The break element has no end tag because it encloses no content. Under XHTML rules, a closing slash (after a space) must be included: <br />. The slash is rarely seen in older HTML documents, so its use will be encouraged but not required.
Attributes: none

<center> ... </center>
    The center element causes displayed text to be centered on the computer screen.
Attributes: none

<em> ... </em>
    This is a "logical element" that will typically cause text to be displayed in italics, but it can be redefined to produce different results in different environments. For most purposes, em and i are interchangeable. See the i element below.
Attributes: none

<font> ... </font>
    The font element controls the appearance of text. The two most commonly used attributes control the size and color of the text.
Attributes: size, color, face

`<hr />` or `<hr>`

    The horizontal rule element draws a shaded horizontal line across the screen. It does not have an end tag. A closing slash (after a space) is required in XHTML. A `noshade` attribute displays the rule as a solid color, rather than shaded.

Attributes: `align`, `color`, `noshade`, `size`, `width`

`<h`*n*`>` ... `</h`*n*`>`

    Up to six levels of headings (for *n* ranging from 1 to 6) can be defined, with decreasing font sizes as *n* increases from 1 to 6.

Attributes: `align`

`<i>` ... `</i>`

    `i` is a "physical element" that forces the included text to be displayed in italics. The actual appearance may depend on the browser and computer used. Compare with the em element above.

Attributes: none

`<img />`

    The `img` element provides a link to an image to be displayed within a document. The image is stored in a separate file, perhaps even at another Web address, the location of which is provided by the `src` attribute.

Attributes: `align`, `border`, `height`, `src`, `vspace`, `width`

`<p>` ... `</p>`

    The p element marks the beginning and end of a paragraph of text content. Note that HTML does not automatically indent paragraphs. Rather, it separates paragraphs with an empty line, with all the text aligned left. It is common to see only the start tag used in HTML documents, without the corresponding end tag. However, the use of the end tag is enforced by XHTML, and this is the style that should be followed.

Attributes: none

`<pre>` ... `</pre>`

    The default behavior of HTML is to collapse multiple spaces, line feeds, and tabs to a single space. This destroys some of the text formatting that you may wish to preserve in a document, such as tabs at the beginning of paragraphs.

    The `pre` element forces HTML to recognize multiple spaces, line feeds, and tabs embedded in text. The default action for `pre` is to use a monospaced font such as `Courier`. This may not always be appropriate, but as line feeds and other text placement conventions are

recognized, `pre` is very useful for embedding programming code examples within an HTML document.
Attributes: none

```
<strong> … </strong>
```
      `strong` is a "logical element" that typically causes text to be displayed in a bold font, but it can be redefined to produce different results in different environments. For most purposes, `b` and `strong` are interchangeable. Compare this with the `b` tag above.
Attributes: none

      Note that most of the elements described here require both start and end tags. The general rule is that any element that encloses content requires both a start and an end tag. The `br` and `hr` elements do not enclose content, so no end tag is needed. However, `br` and `hr` should include a closing slash in their tags in order to be XHTML-compatible—for example, `<br />` rather than `<br>`, with a space before the slash.

Description of attributes:
These descriptions may not include all possible values. For a complete listing, consult an HTML reference manual.

```
align = "…"
```
Values: `"left"`, `"right"`, or `"center"`
Aligns text horizontally.

```
background = "…"
```
Value: the URL of a gif- or jpeg-format graphics file
      Setting the background attribute displays the specified image as the background behind a displayed HTML document page. Depending on the image size (in pixels), background images may automatically be "tiled," resulting in a repeating image that can be visually distracting. It is not necessary to use background images, and they should be used with care.

```
bgcolor = "…"
```
Values: Background colors can be set either by name or by specifying the intensity of the red, green, and blue colors. This topic is addressed in Section 2.5.

```
border="…"
```
Value: The width, in pixels, of a border surrounding an image

```
color = "…"
```
Values: Text colors can be set either by name or by directly specifying the intensity of the red, green, and blue colors. See Section 2.5.

face = "..."
Values: Font typefaces can be set either generically, with cursive, monospace, sans-serif, or serif, or with specific font names supported by the user's computer.

The generic names should always produce something that looks reasonable on any computer, but specific font names that are not available on the user's computer may produce unexpected results.

height = "..."
Value: The height, in pixels, of an image.

href = "..."
Value: The URL of an external or internal Web resource or the name of an internal document reference.

hspace = "..."
Value: The horizontal space, in pixels, between an image and the surrounding text.

name = "..."
Value: The name assigned to an internal document reference through an "a" element.

size = "..."
Values: An unsigned integer from 1 to 7 or a signed number from +1 to +6 or –1 to –6.

An unsigned integer is an absolute font size, which may be system-dependent. The default value is 3. A signed integer is a font size relative to the current font size, larger for positive values and smaller for negative values.

For the hr element, size is the vertical height of the horizontal rule, in pixels.

src = "..."
Value: The URL of a graphics file. For local use, images and their HTML document are usually stored in the same folder.

text = "..."
Values: The text attribute, used with the body element, selects the color of text in a document, which prevails unless overridden by a font attribute.

vspace = "..."
Value: The vertical space, in pixels, between an image and the surrounding text.

width = "..."
Values: The width of an image or horizontal rule, in pixels or as a percent of total screen width. For example, width="80" is interpreted as a width of 80 pixels, but width="80%" is a width equal to 80 percent of the total screen width.

Document 2.1 illustrates how some of these elements are used.

Document 2.1 (tagExamples.htm)

```html
<html>
<head>
<title>Tag Examples</title>
</head>
<body bgcolor="white">
<h1>Here is a Level 1 Heading</h1>
<h2>Here is a Level 2 Heading</h2>
<hr />
<pre>
        Here is some <strong><em>preformatted
text</em></strong> that has
        been created with the pre element. Note that it
retains the
paragraph tab
included
in the <b><i>original        document</b></i>. Also, it does
not "collapse" line feeds
and
            white            spaces. Often, it is easier to
use preformatted text than it
is to use markup to get the same effect. Note, however, that
the default
rendering of
preformatted text is to use a monospaced Courier font. This
is often a good choice for
displaying code in an HTML document, but perhaps not a good
choice for other kinds of text content.
</pre><p><center>
<img src="checkmark.gif" align="left" />Here, a small
graphic (the check box) has been inserted into
the document using the "img" element. This text is outside
the preformatted
region, so the default font is different. If you look at the
original document, you can also see that
white          spaces and line    feeds are now collapsed.
</p><p>
Note too, that the text is now centered. The way the text is
displayed will
```

```
depend on how you
have the display window set in your browser. It may change
when you go from full screen to a window, for example.
</center></p><p>
Centering is now turned off. The default text alignment is
to the left of your screen.
You can change the size and color of text <font size="7"
color="blue"> by using the &lt;font&gt;</font>
<font color="purple">element.</font>
</body>
</html>
```

Below is one rendering of Document 2.1. The small checkbox graphic has been created with the Windows Paint program. The actual text displayed in your browser is larger than this, but the output image has been reduced in size (perhaps to the extent of not being readable) to fit on the page. Moreover, because of the line feeds imposed on the text of this code example by the page width, the output looks a little different from what you might expect. So, you have to try this document on your own browser.

Here is a Level 1 Heading

Here is a Level 2 Heading

Here is some *preformatted text* that has
been created with the pre element. Note that it retains the
paragraph tab
included
in the *original*   *document*. Also, it does not "collapse" line feeds
and
          white   spaces. Often, it is easier to use preformatted text than it
is to use markup to get the same effect. Note, however, that the default rendering of
preformatted text is to use a monospaced Courier font. This is often a good choice for
displaying code in an HTML document, but perhaps not a good choice for other kinds of text content.

 Here, a small graphic (the check box) has been inserted into the document using the "img" element. This text is outside the preformatted region, so the default
font is different. If you look at the original document, you can also see that white spaces and line feeds are now collapsed.

Note too, that the text is now centered. The way the text is displayed will depend on how you have the display window set in your browser. It may change when you
go from full screen to a window, for example.

Centering is now turned off. The default text alignment is to the left of your screen. You can change the size and color of text by using the

<font> element.

Document 2.1 answers an interesting question: How can HTML display characters that already have a special meaning in the HTML language or that do not appear on the keyboard? The angle brackets (< and >) are two such characters because they are part of HTML tags. They can be displayed with the < and > escape sequences (for the "less than" and "greater than" symbols from mathematics). There are many standardized escape sequences for special symbols. A list of some of them is given in Appendix 2.

2.2 HTML Syntax and Style

A general characteristic of programming languages is that they have very strict syntax rules. HTML is different in that regard, as it is not highly standardized. The positive spin on this situation is to call HTML an "open standard," which means that self-described bearers of the standard can treat the language as they see fit, subject only to usefulness and market acceptance. HTML has an established syntax, but it is very forgiving about how that syntax is used. For example, when a browser encounters HTML code that it does not understand, typically it just ignores it rather than crashing, as a "real" program would do.

Fortunately, market forces—the desire to have as many people as possible accept your browser's interpretation of HTML documents—have forced uniformity on a large subset of HTML. This book adopts some HTML style conventions and syntax that are as platform-independent as possible. Although these "rules" might seem troublesome if you are not used to writing stylistically consistent HTML documents, they should actually help beginners by providing a more stable and predictable working environment. The only things worse than having syntax and style rules are having no rules or rules that nobody follows.

Some of the style rules used in this book are listed below. Under the circumstances of HTML, they are more accurately referred to as "guidelines." Some of them will make more sense later on, as you create more complicated documents.

1. Spell the names of HTML elements in lowercase letters.

Unlike JavaScript and some other languages, the HTML language is not sensitive to case. Thus, `<html>`, `<HTML>`, and `<hTmL>` are equivalent. However, the XHTML standard requires element names to be spelled with lowercase letters. In the earlier days of HTML, many programmers adopted the style of using uppercase letters for element names because they stood out in a document. You will often still see this style in Web documents. Nonetheless, we will consistently use lowercase letters for element names.

2. Use the `pre` element to enforce text layout whenever it is reasonable to use a monospaced font (such as `Courier`).

HTML always collapses multiple "white space" characters— spaces, tabs, and line breaks—into a single space when text is displayed. The easiest way to retain white space characters is to use the `pre` element. Other approaches may be needed if proportional fonts are required. Furthermore, tabbed text may still not line up, as different browsers have different default settings for tabs.

3. Nest elements properly.

Improperly nested elements can cause interpretation problems for your browser. Even when browsers do not complain about improperly nested elements, HTML is easier to learn, read, and edit when these restrictions are enforced.

Recall the following markup in Document 2.1:

```
Here is some <strong><em>preformatted
text</em></strong>
```

If you write this as

```
Here is some
<strong>
        <em>
                ...{text}
        </em>
</strong>
```

it is easy to see that the em element is properly nested inside the strong element. If this is changed to

```
<strong><em> ...{text} </strong></em>
```

your browser probably will not complain, but it is not good programming style.

4. Enclose the values of attributes in single or double quotes.

In Document 2.1, bgcolor="white" is an attribute of <body>. Browsers generally will accept bgcolor=white, but the XHTML standard enforces the use of quoted attribute values. This book is consistent about using double quotes unless attribute values appear inside a string that is surrounded with double quotes (for example, an attribute value embedded in a parameter in the document.write() method). Then attribute values will be single-quoted.

2.3 Using the script Element

The script element usually (but not always) appears inside the head element, after the title element. Following is a description of script along with its essential attributes:

```
<script language="javascript" type="text/javascript">
    ...
```

```
</script>
```
Attributes: language, type, src

The values usually assigned to the language and type attributes are language="javascript" and type="text/javascript". The values shown in the description are default values, so for documents using JavaScript, inclusion of these attributes is usually not actually required.

The src attribute has a value corresponding to the name of a file containing JavaScript script, usually (but not necessarily) with a .js extension. This attribute is used in a later chapter.

2.4 Creating and Organizing a Web Site

Obviously this is a major topic, a thorough investigation of which would go far beyond the reach of this text. There is an entire industry devoted to hosting and creating Web sites, including helping a user obtain a domain name, providing storage space, developing content, and tracking access. For the purposes of a course based on this text, the goal is extremely simple: create a Web site sufficient to display the results of work done during the course.

The first step toward creating a Web site is establishing its location. In an academic environment, a college, university, or department computer may provide space for web pages. A URL might look something like this:

http://www.myuniversity.edu/~username

where the "~" symbol indicates a directory where Web pages are stored. Together with a user name, this URL directs a browser to the home Web directory for that user. As noted in Chapter 1, as HTML documents are not automatically Internet-accessible, your Web pages for this book may be accessible only locally on your own computer.

In this home directory there should be at least one file called index.htm (or index.html). UNIX systems favor the .html extension, but Windows users should use the three-character .htm extension to remain compatible with Windows file extension conventions. This is the file that will open automatically in response to entering the above URL. That is, the index.htm file is the "home page" for the Web site. This home page file could be named something else, but then its name would have to be added to the URL:

http://www.myuniversity.edu/~username/HomePage.htm

An `index.htm` file can contain both its own content as well as links to other content (hyperlinks), including other pages on the user's Web site and to external URLs. Following are four important kinds of links:

1. Links to other sites on the World Wide Web.
    The following is the basic format for globally linking Web pages:

syntax: <a href=*"{URL of Web page}"*>
        {description of linked Web page}</a>

The URL may refer to a completely different Web site, or it may be a link to documents in the current folder or a subfolder within that folder.

2. Links to images.
    The `img` element is used to load images for display or to use as a page background:

syntax: <img src=" *{URL plus image name}*" align="..."
        height="..." width="..." />

The image may exist locally or it may be at a different Web site. The `align`, `height`, and `width` attributes, which can be used to position and size an image, are optional. However, for high-resolution images, it is almost always necessary to specify the height and width as a percentage of the full page or as a number of pixels in order to reduce the image to a manageable size in the context of the rest of the page. Resizing the image, if possible, will solve this problem.
    You can also make a "clickable image" to direct the user to another link:

Syntax: <a href="*{URL of Web page}*">
        <img src=" *{URL plus image name}*" align="..."
        height="..." width="..." /></a>

3. Links to e-mail addresses.
    An e-mail link is an essential feature that allows users to communicate with the author of a Web page.

syntax: <a href="mailto:*{e-mail address}*">
        {description of recipient}</a>

Often, but not necessarily, the *{description of recipient}* is also the e-mail address. The actual sending of an e-mail is handled by the default mailer on the sender's computer.

4. Internal links within a document.

Within a large document, it is often convenient to be able to move from place to place within the document using internal links.

Syntax: <a href=" *{#internal link name}*">
    {description of target position}</a>
    ...
    <a name=" *{internal link name}*">*{target text}*</a>

The "#" symbol is required when specifying the value of the href attribute, in order to differentiate this internal link from a link to another (external) document.

The careless use and specification of hyperlinks can make Web sites very difficult to maintain and modify. As noted above, every Web site should have a "home" directory containing an index.htm file. In order to make a site easy to transport from one computer to another, all other content should be contained either in the home directory or in folders created within that directory. References to folders that are not related in this way should be avoided, as they will typically have to be renamed if the site is moved to a different computer. Although it is allowed as a matter of syntax to give a complete (absolute) URL for a local Web page, this should be avoided in favor of a reference relative to the current folder.

This matter is important enough to warrant a complete example. Document 2.2a–c shows a simple Web site with a home folder on a Windows desktop called home and two subfolders within the home folder named homework and personal. Each subfolder contains a single HTML document, homework.htm in homework and resume.htm in personal.

Document 2.2a (index.htm)

```html
<html>
<head>
<title>My Page</title>
</head>
<body>
<!-- These absolute links are a bad idea! -->
Here are links to
<a href="C:/Documents and Settings/David/desktop/
JavaScript/Book/homework.htm">homework</a> and
<a href="C:/Documents and Settings/
```

```
  David/desktop/JavaScript/Book/resume.htm">
personal documents.</a>
</body>
</html>
```

Document 2.2b (resume.htm)

```
<html>
<head>
<title>Resumé</title>
</head>
<body>
Here is my resumé.
</body>
</html>
```

Document 2.2c (homework.htm)

```
<html><head>
<title>Homework</title>
</head>
<body>
Here are my homework problems.
</body>
</html>
```

Note that Document 2.2a uses forward slashes to separate the directories and file names. This is consistent with UNIX syntax, but Windows/DOS systems use backward slashes. Forward slashes are the HTML standard, and they should always be used even though backward slashes may also work. Another point of interest is that UNIX directory paths and filenames are case-sensitive, but Windows/DOS paths and filenames are not. This could cause problems if you develop a Web page on a Windows/DOS computer and then move it to a UNIX-based system. As a matter of style, you should be consistent about case in directory and file names even when it appears not to matter.

Absolute references to a folder on a particular Windows computer desktop are a bad idea because such references will have to be changed if the index.htm file is moved to a different place on the same computer, or to a different computer—for example, to a UNIX university department computer with a different directory/folder structure. Document 2.2d shows the preferred solution. Now the paths to homework.htm and resume.htm are given relative to the home folder, wherever the index2.htm file resides. (Remember that this file, no longer named index.htm, will not be recognized as a default home page.) This document assumes that folders homework and personal exist in the home folder. The relative URL should work without modification when the Web site is moved to a different computer. If the Web site is moved, only a single reference, the one to the `index2.htm` file, has to be changed.

Document 2.2d (`index2.htm`, a new version of `index.htm`)

```html
<html>
<head>
<title>My Page</title>
</head>
<body>
<!-- Use these relative links instead! -->
Here are links to
<a href="homework/homework.htm">homework</a>
and <a href="personal/resume.htm">personal documents.</a>
</body>
</html>
```

When designing a Web site proper attention to the use of relative URLs from the very beginning will save a lot of time in the future!

2.5 Selecting and Using Colors

As previously noted, several attributes, such as `bgcolor`, are used to set colors of text or backgrounds. Colors may be identified by name or by a six-character hexadecimal numeric code that specifies the strength of the signal emitted from the red, green, and blue electron "guns" that excite the corresponding phosphors on a cathode ray tube color monitor screen. This convention is retained even when other display technologies are used. The **hex code** is in the format `#RRGGBB`, where each color value can range from `00` (turned off) to `FF` (maximum intensity).

There are many color names in use on the Web, but only 16 are standardized, representing the 16 colors recognized by the Windows VGA color palette.

Table 2.1. A list of 16 standard HTML color names and hex hodes

Color Name	Hexadecimal Code
aqua	#00FFFF
black	#000000
blue	#0000FF
fuchsia	#FF00FF
gray	#808080
green	#008000
lime	#00FF00
maroon	#800000
navy	#000080
olive	#808000
purple	#800080
red	#FF0000
silver	#C0C0C0
teal	#008080
white	#FFFFFF
yellow	#FFFF00

These colors are listed in Table 2.1. The problem with additional color names is that there is no enforced standard for how browsers should interpret them. Two examples: magenta probably should be, but does not have to be, the same as fuchsia; ivory is a nonstandard color that should be rendered as a yellowish off-white. The colors in Table 2.1 are standardized in the sense that all browsers should associate these 16 names with the same hexadecimal code. Of course, variations can still occur because monitors themselves respond somewhat differently to the same name or hex code; blue on my computer monitor may look somewhat different than blue on your monitor.

Note that the standardized colors use a limited range of hex codes. With the exception of silver (nothing more than a lighter gray), the RGB gun colors are off (00), on (FF), or halfway on (80).

What should you do about choosing colors? Favor standardized colors, and if you wish to make an exception, try it in as many browser environments as possible. Be careful to choose background and text colors so that the text will always be visible against its background. The safest approach for setting colors in the body element is to specify both background and text colors. This will ensure that default colors set in a user's browser will not result in unreadable text.

If you are not sure whether a color name is supported and what it looks like on your monitor, you have nothing to lose by trying it. If you set bgcolor="lightblue", you will either like the result or not. If a color name is not recognized by your browser, the result will be unpredictable, but not catastrophic. There are (of course) numerous Web sites that can help you work with colors, including getting the desired result with hex codes.

2.6 Using Cascading Style Sheets

As you create more Web pages, you may wish to impose a consistent look for all of your pages or for groups of related pages. It is tedious to insert elements for all the characteristics you may wish to replicate—font size, font color, background color, and so forth. Style sheets make it much easier to replicate layout information in multiple documents..A complete discussion of style sheets is far beyond the scope of this book, as there are many different kinds of style sheets, many ways to make use of them, and many browser-specific nuances. This book uses **cascading style sheets** (CSSs), which are widely accepted as a default kind of style sheet, but presents only a *small* subset of all the possibilities! By way of introduction, Document 2.3 illustrates the use of a style element to establish the default appearance of the body of an HTML document.

Document 2.3 (`style1.htm`)

```
<html>
<head>
<title>Style Sheets</title>
<style title="David's default" type="text/css">
       body.bright {background: red; font: 16pt serif;
          color: blue; font-style: italic; font-weight: bold}
</style>
</head>
<body class="bright">
Here is the body.
</body>
</html>
```

The `style` element has an optional `title` attribute and a `type` attribute set equal to `"text/css"`, where the `css` stands for cascading style sheet. This `style` element gives the `body` style a name (`bright`) and sets the document background color to red and the default font to bold, 16-point serif, blue, and italicized. Note the use of the dot notation to assign a **class name** to the **style rule(s)** established for the element, and the use of the name later (`class="bright"`) with the `class` attribute in the `<body>` tag. Each style rule is terminated with a semicolon. So, for example, the line

```
{font: 16pt serif; color: blue;}
```

gives one rule for setting font properties and a second for setting text color. When multiple properties are set for the same element, they are enclosed in curly brackets.

   For this simple example, with styles applying only to a single `body` element, the class name is optional. In general, several different style rules can apply to the same HTML element. For example, several different style rules could be established for paragraphs (`<p>` ... `</p>`), each of which would have its own class name.

   In summary, style specifications follow a hierarchy:

`style` element → other HTML elements*[.class name]* →
                 properties → value(s)

where the class name (without the brackets) is optional.

   How did CSSs get that name? The answer is that the properties set for an element cascade down, or are "inherited," by other elements contained within it unless those elements are assigned their own style properties. So, for example, properties set for the `body` element are inherited by the p and h1 elements because these are contained within the body element. Properties set for the head element are inherited by content appearing in the title element.

CSSs can be used to modify the appearance of any HTML element that encloses content. Following are some properties that can be specified in style sheets.

Background properties

background-color

When used in a body element, background-color sets the background color for an entire document. It can also be used to highlight a paragraph, for example, when used with a p element.

background-image

This property is used with a URL to select an image file (gif or jpeg) that will appear as a background. Typically, this is used with a body element, but it can also be used with other elements, such as p. For other background properties that can be used to control the appearance of a background image, consult an HTML reference text.

background

This allows you to set all background properties in a single rule.

Color property

The color property sets the default color for text, using the descriptions discussed in Section 2.5.

Font properties

font-family

Font support is not completely standardized. However, browsers that support style sheets should support at least the generic font families listed in Table 2.2.

Table 2.2. Generic font families

Generic Name	Example
cursive	Zapf-Chancery
monospace	Courier
sans-serif	Arial
serif	Times

Example: font-family: Arial, sans-serif;

font-size

This property allows you to set the actual or relative size of text. You can use relative values, such as large, small, larger, smaller (relative to a default size); a percentage, such as 200% of the default size;

or an actual point size such as 16pt. Some sources advise against using absolute point sizes because a point size that is perfectly readable on one system might be uncomfortably small on another. For our purposes, specifying the point size is probably the easiest choice.

Example: font-size: 24pt;

font-style

> This property allows you to specify normal, italic, or oblique fonts.

Example: font-style: italic;

font-weight

> This property allows you to select the font weight. You can use values in the range from 100 (extra light) to 900 (extra bold), or words: extra-light, light, demi-light, medium, demi-bold, bold, and extra-bold. Some choices may not have a noticeable effect on some fonts in some browsers.

Example: font-weight: 900;

font

> This property allows you to set all font properties within one style rule.

Example: font: italic 18pt Helvetica, sans-serif;

How will your browser interpret a generic font name? For the generic name serif, it will pick the primary serif font that it supports—probably Times or Times Roman. Browsers will probably also recognize specific font names such as Times or Helvetica (a sans-serif font). If you specify a font name not supported by your browser, it will simply ignore your choice and use its default font for text. It is possible to list several fonts, in which case your browser will select the first one it supports. For example, consider this rule:

font-family: Arial, Helvetica, sans-serif;

Your browser will use an Arial font if it supports that, Helvetica if it does not support Arial but does support Helvetica, or, finally, whatever sans-serif font it does support. By giving your browser choices, with the generic name as the last choice, you can be reasonably sure that text will be displayed with a sans-serif font.

Text properties
Of the many text properties, just three that may be useful are shown below.

`text-align`
  This is used in block elements such as `p`. It is similar in effect to the HTML `align` attribute. The choices are `left`, `right`, `center`, and `justify`. With large font sizes, `justify` may produce odd-looking results.

Example: `text-align: center;`

`text-indent`
  Recall that paragraphs created with the `p` element do not indent the first word in the paragraph. (HTML inserts a blank line, but left-justifies the text.) This property allows you to set indentation using typesetting notation or actual measurements. I suggest the use of actual English or metric measurements—inches (`in`), millimeters (`mm`), or centimeters (`cm`).

Example: `text-indent: 0.5in;`

`white-space`
  The value of this property is that you can prevent spaces from being ignored. (Remember that the default HTML behavior is to collapse multiple spaces and other nonprintable characters into a single blank space.) You can use the HTML `pre` element by itself, instead, but this causes the text to be displayed in a monospaced font such as Courier. (At the time this book was written, not all browsers supported this property.) The example given here retains white space regardless of the typeface being used.

Example: `white-space: pre;`

  Styles are not restricted just to the body element. For example, paragraphs (`<p>` ... `</p>`) and headings (`<hn >` ... `</hn>`) can also have styles associated with them. You can also set styles in selected portions of text using the `span` element, and in blocks of text using the `div` element.

`<div>` ... `</div>`
Attributes: `align`, `style`

`<span>` ... `</span>`
Attributes: `align`, `style`
Values for align: `"left"` (default), `"right"`, `"center"`

You can create style sheets as separate files and then utilize them whenever you wish to use a particular style on a Web page. This makes it easy to impose a uniform appearance on multiple Web pages. Documents 2.4a and 2.4b show a simple example.

Document 2.4a (body.css)

```
body {background:silver; color:white; font:24pt Times}
h1 {color:red; font:18pt Impact;}
h2 {color:blue; font:16pt Courier;}
```

Document 2.4b (style2.htm)

```
<html>
<head>
<title>Style Sheet Example</title>
<link href="body.css" rel="stylesheet"
   type="text/css" />
</head>
<body>

   <h1>Heading 1</h1>
   <h2>Heading 2</h2>
   Here is some text.
</body>
</html>
```

This example shows *(See Color Example 3 for full-color output.)* how to create a file, body.css, containing style elements that can be applied to any document by using the link element, as in Document 2.4b. The .css extension is standard, but not required. (You could use .txt, for example.) Although this example is very simple, the concept is powerful because it makes it easy to create a standard style for all your documents that can be invoked with the link element. The Impact font chosen for h1 headings will not be supported by all browsers, in which case the default font will be used in its place.

The attributes of link include href, which contains the URL of the style sheet file, the rel="stylesheet" (relationship) attribute, which describes how to use the file (as a style sheet), and the type, which should be "text/css", just as it would be defined if you created a style element directly in the head element. In this example, body.css is in the same folder as style2.htm. If you keep all your style sheets in a separate folder, you will need a more explicit URL.

It is worth re-emphasizing that this discussion of style sheets has barely scratched the surface of the subject. Style sheets can make your Web pages more visually appealing and can greatly simplify your work on large Web projects. Some Web developers advocate replacing *all* individual formatting elements, such as font and its attributes, with style sheet specifications. In newer versions of HTML, and in XHTML, the use of individual formatting elements is "deprecated," but there is little likelihood that support for them will disappear from browsers in the foreseeable future. A course based on this book does not require the use of cascading style sheets unless it is asked for specifically.

2.7 Another Example

Documents 2.5a and 2.5b show how to use a style sheet file to specify different background and text colors for different sections of text.

Document 2.5a (rwb.css)

```
p.red {background:red;color:blue;font:20pt Times}
div.white {background:white;color:red;font:20pt Times}
span.blue {background:blue;color:white;font:20pt Times}
```

DOCUMENT 2.5b (rwb.htm)

```
<html>
<head>
<title>A Red, White, and Blue Document</title>
<link href="rwb.css" rel="stylesheet" type="text/css" />
</head>
<body>
<img src="stars.jpg" height="150" width="250" />
<p class="red">
This text should be blue on a red background.
</p><p><div class="white" style="font-style: italic;">
This text should be red on a white background.
</div></p>
<p><span class="blue">This text should be white on a blue
background.</span>
</p>
</body>
</html>
```

This text should be blue on a red background

This text should be red on a white background

This text should be white on a blue background

(See Color Example 4 for full-color output.)

The stars (which are supposed to be red, silver, and blue) have been drawn using the Windows Paint program.

3. HTML Tables, Forms, and Lists

Chapter 3 explains how to create HTML tables, forms, and lists; how to organize documents for user input by combining forms and tables; and how to send the contents of a form back to its creator.

3.1 The `table` Element

3.1.1 Basic Table Formatting

HTML **tables** and **forms** are the two most important ways to organize the content of a Web page. Forms are critical because they provide a user interface for JavaScript. It is sometimes helpful to organize information in a form through the use of one or more tables. With that approach in mind, we first consider tables.

Since HTML ignores text formatting, such as white space and line feeds (**Enter**), it can be difficult to control the placement of content on a web page, and the addition of graphics only compounds the problem. An easy way to gain some control is to create a table, using the `table` element. Then the relative locations of text and graphics can be established by entering them into cells of the table. Within the start and end tags, `<table>` ... `</table>`, rows and cells are defined with the `tr` ("table row") and `td` ("table data") elements, which are nested as follows:

```
<table>
      <tr>
            <td> ... </td>  {as many columns as you need...}
            ...
      </tr>
      {as many rows as you need...}
      ...
</table>
```

The <tr> ... </tr> tags define the rows and the <td> ... </td> tags define cells in columns within those rows. You can define as many rows and columns as you need. With these elements, you can organize information in a familiar spreadsheet-like row-and-column format. Document 3.1 shows how to use a table to organize and display some results from residential radon testing.

Document 3.1 (radonTable.htm)

```
<html>
<head>
<title>Radon Table</title>
</head>
<body>
<h1>Results of radon testing</h1>
<p>
The table below shows some radon levels measured in resi-
dences.<br /> For values greater than or equal to 4 pCi/L,
action should be taken<br /> to reduce the concentration of
radon gas. For values greater than or<br />
equal to 3 pCi/L, retesting is recommended.
</p>
<table>
  <tr bgcolor="silver">
    <td>Location</td><td>Value, pCi/L</td>
  <td>Comments</td></tr>
  <tr>
    <td>DB's house, basement</td><td>15.6</td>
    <td bgcolor="pink">Action should be taken!</td></tr>
  <tr>
    <td>ID's house, 2nd floor bedroom</td><td>3.7</td>
    <td bgcolor="yellow">Should be retested.</td></tr>
  <tr>
    <td> FJ's house, 1st floor living room</td><td> 0.9</td>
    <td bgcolor="lightgreen">No action required.</td></tr>
  <tr>
    <td> MB's house, 2nd floor bedroom</td><td>2.9</td>
    <td bgcolor="lightgreen">No action required.</td></tr>
</table>
</body>
</html>
```

Results of radon testing

The table below shows some radon levels measured in residences. For values greater than or equal to 4 pCi/L, action should be taken to reduce the concentration of radon gas. For values greater than or equal to 3 pCi/L, retesting is recommended.

Location	Value, pCi/L	Comments
DB's house, basement	15.6	Action should be taken!
ID's house, 2nd floor bedroom	3.7	Should be retested.
FJ's house, 1st floor living room	0.9	No action required.
MB's house, 2nd floor bedroom	2.9	No action required.

(See Color Example 5 for full-color output.)

The syntax for tables includes several possibilities in addition to `tr` and `td` for customizing the appearance of a table, including the `caption` element, which associates a caption with the table, and the `th` element, which is used to create a "header" row in a table by automatically displaying text in bold font. (The `th` element can be used anywhere in a table in place of `td`.) The `caption`, `td`, `th`, and `tr` elements are used only inside the start and end tags of a `table` element: `<table> … </table>`. With these elements, a more comprehensive table layout looks like this:

```
<table>
      <caption> … </caption>
      <tr>
      <!-- Use of th in place of td is optional. -->
            <th> … </th>
            …
      </tr>
      <tr>
            <td> … </td>
            …
      </tr>
      …
</table>
```

The attributes associated with these tags all have default values, so you do not have to give them values. You can create a table without using any attributes at all and then add attributes as needed. In Document 3.1, the only specified attribute is the background color in some cells. An easy way to familiarize yourself with the effects of specifying table attributes and their values is to experiment with Document 3.1.

3.1.2 Merging Cells across Rows and Columns

If you are familiar with creating tables in a word processing application, you know that it is easy to create more complicated table layouts by merging cells across rows and columns. You can also do this with HTML forms, using the `colspan` and `rowspan` attributes. Document 3.2 shows a table that displays cloud names, altitudes, and indicates whether or not they produce precipitation.

Document 3.2 (`cloudType.htm`)

```
<html>
<head>
```

```
<title>Cloud Type Chart</title>
</head>
<body>
<table border="2">
<caption>Cloud Type Chart</caption>
<tr>
   <th align="center">Altitude</th>
   <th colspan="2">Cloud Name</th></tr>
<tr><td align="center" rowspan="3">High</td>
     <td colspan="2">Cirrus</td></tr>
     <tr><td colspan="2">Cirrocumulus</td></tr>
     <tr><td colspan="2">Cirrostratus</td></tr></tr>
<tr><td align="center" rowspan="2">Middle</td>
     <td colspan="2">Altocumulus</td></tr>
     <tr><td colspan="2">Altostratus</td></tr></tr>
<tr><td align="center" rowspan="5">Low</td>
     <td>Cumulus</td>
     <td>nonprecipitating</td></tr>
<tr><td>Altocumulus</td>
     <td>nonprecipitating</td></tr>
<tr><td>Stratocumulus</td>
     <td>nonprecipitating</td></tr>
     <tr><td>Cumulonimbus</td>
     <td align="center"
        bgcolor="silver">precipitating</td></tr>
     <tr><td>Nimbostratus</td> <td align="center"
        bgcolor="silver">precipitating</td></tr></tr>
</table>
</body></html>
```

It is much more tedious to merge cells across rows in columns in an HTML table than it is in a word processing program, so you have to plan your table in advance. Even then, you should be prepared for some trial-and-error editing!

A summary of some table-related elements and their attributes is given below. All the elements except `table` itself should appear only inside a `table` element.

Cloud Type Chart

Altitude	Cloud Name	
High	Cirrus	
	Cirrocumulus	
	Cirrostratus	
Middle	Altocumulus	
	Altostratus	
Low	Cumulus	nonprecipitating
	Altocumulus	nonprecipitating
	Stratocumulus	nonprecipitating
	Cumulonimbus	precipitating
	Nimbostratus	precipitating

`<caption> ... </caption>`
Attributes: align
  Displays the specified text as a caption for a table. Earlier versions of HTML support only "top" (the default value) or "bottom" for the value of the align attribute. Some browsers may allow "center" as a value for align, which is worth noting because this might often be the alignment of choice for a table caption.

`<table> ... </table>`
Attributes: border, bordercolor, cellpadding, cellspacing, width
  Contains table-related and other elements.

`<td> ... </td>`
Attributes: align, bgcolor, colspan, nowrap, rowspan, width
  Does not contain other table-related elements.

`<th> ... </th>`
Attributes: align, bgcolor, colspan, nowrap, rowspan, valign, width
  The th element works just like the td element except it automatically displays text in bold font, serving as headings for table columns. It does not contain other elements.

`<tr> ... </tr>`
Attributes: align, bgcolor, valign
  Contains td or th elements.

Description of attributes:

align = "..."
Values: "left", "right", or "center"
  Aligns text horizontally. When align is specified in a tr element, its value will be overridden if it is specified again within a td element in that row.

bgcolor = "..."
Values: color names or hexadecimal values "#*RRGGBB*"
  Sets the background color for a cell or row. When bgcolor is specified in a tr element, its value will be overridden if it is specified again within a td element in that row.

`border = "..."`
Values: an integer number of pixels
Adds a border to the table and its cells. A value is optional. If it is included, an additional colored border is added around the outer boundary of the table.

`bordercolor = "..."`
Values: color names or hexadecimal values `"#RRGGBB"`
Sets the color of a table border.

`cellpadding = "..."`
Values: an integer number of pixels
Defines vertical spacing between cells in a table.

`cellspacing = "..."`
Values: an integer number of pixels
Defines horizontal spacing between cells in a table.

`colspan = "..."`
Values: an integer
Defines how many columns a cell will span.

`nowrap`
Prevents text from being automatically wrapped within a cell. It does not have a value.

`rowspan = "..."`
Values: an integer
Defines how many rows a cell will span.

`valign = "..."`
Values: `"top"`, `"middle"`, or `"bottom"`
Aligns text vertically. When `valign` is specified in a `tr` element, its value will be overridden if it is specified again within a `td` element in that row.

`width = "..."`
Values: a number or a percentage
Specifies table or cell width in pixels (`width="140"`) or as a percentage of the window or table header width (`width="80%"`).

3.2 The `form` **Element**

One of the most important applications of HTML documents is to provide the Web page equivalent of a paper form. In some cases, a form just helps to organize user input to a Web page. Often, an online form includes provisions for sending a completed form back to the author of the Web page. In other cases, the form may act as an I/O interface in which a user provides input and the Web page provides results from calculations or other actions. This use of forms is especially important for the material presented in later chapters of this book.

HTML forms are defined by the `form` element, using start and end tags: `<form>` ... `</form>` tags. The attributes of the `form` element are:

`action = "..."`
Value: a programmer-supplied URL that identifies a processing script or `mailto:` followed by an e-mail address. In this book, I always use the `mailto:` action. For example,
`action="mailto:my_mail@my_univ.edu"`.

`enctype="..."`
Value: In this book, I use only `enctype="text/plain"`. In combination with `method="post"`, this will transmit form data with the name of the form field followed by an "=" sign and the value of the field, which makes it easy to interpret the contents of a form that has been submitted.

`method = "..."`
Values: `"get"`, `"post"`
The `method` attribute controls how data from a form is sent to the URL or e-mail address identified in the `action` attribute. In this book, I use the `"post"` value because it is the easiest way to transmit form data in an easily readable format.

`name = "..."`
Value: a programmer-selected name that is used to identify the form.
The `name` attribute is needed only if a document contains more than one form.

Forms contain one or more input **fields** identified by <input /> tags. As the input element does not enclose content, it has no end tag, so it requires a closing slash for XHTML compliance. The most important attribute of input is its type. There are several field types that have well-defined default behaviors in HTML. The possible values are listed in Table 3.1.

Table 3.1. Values for the input element's type attribute

Field Type	Description
type = "button"	Provides a programmer-defined action to be associated with the field through the use of an event handler such as onclick.
type = "checkbox"	Allows selection of one or more values from a set of possible values.
type = "hidden"	Allows the definition of text fields that can be accessed by a JavaScript script but are not displayed in a document.
type = "password"	Allows entry of character data but displays only asterisks.
type = "radio"	Allows selection of one and only one value from a set of possible values.
type = "reset"	Used to reset all form fields to their default values.
type = "submit"	Processes form contents according to a specified method and action.
type = "text"	Allows entry of character data.

There is no field type specifically for numerical values. This will be significant when we start to use JavaScript to process the contents of forms. The use of event handlers, mentioned in the description of the "button" field type, is discussed in Chapters 4 and 6.

Following is a list of attributes for the input element:

checked
Value: none
        Applies to type="radio" and type="checkbox" only.

maxlength="..."
Value: Maximum number of characters that can be entered in the field. This value can be greater than the value given for the size attribute.

`name="..."`
Value: A programmer-supplied name for the field. The name should follow the variable-naming conventions for JavaScript (see Chapter 4) in order to facilitate its use in JavaScript scripts.

`readonly`
Value: none
     Prevents field values in `type="text"` or `text="password"` from being changed.

`size="..."`
Value: width of the displayed field, in characters.

`type="..."`
Values: See Table 3.1.

`value="..."`
Value: a programmer-supplied default value that will be displayed in the field. This value can be overridden by user input unless the `readonly` attribute is also specified.

The `form` element typically contains a combination of document text and input fields. The text can be used to explain to the user of the form what kind of input is expected. Document 3.3 illustrates a simple example that uses several input field types.

Document 3.3 (`location.htm`)

```html
<html>
<head>
<title>Data Reporting Site Information</title>
</head>
<body>
<form>
  Please enter your last name:
  <input type="text" name="last_name" size="20"
    maxlength="20" /><br />
  Please enter your latitude:
  <input type="text" name="lat" value="40" size="7"
    maxlength="7" />
   N <input type="radio" name="NS" value="N" checked />
    or S <input type="radio" name="NS" value="S" /><br />
  Please enter your longitude:
  <input type="text" name="lon" value="75" size="8"
    maxlength="8" />
    E <input type="radio" name="EW" value="E" /> or W
```

```
    <input type="radio" name="EW" value="W" checked /><br />
    Please enter your elevation:
    <input type="text" name="elevation" size="8" maxlength="8"
    /> meters<br />
    Please indicate the seasons during which your site reports
      data:<br />
    Winter: <input type="checkbox" name="seasons"
      value="Winter" />
    Spring: <input type="checkbox" name="seasons"
      value="Spring" />
    Summer: <input type="checkbox" name="seasons"
      value="Spring" />
    Fall: <input type="checkbox" name="seasons"
      value="Fall" />
</form>
</body>
</html>
```

| Please enter your last name: |
| Please enter your latitude: 40   N ⊙ or S ○ |
| Please enter your longitude: 75   E ○ or W ⊙ |
| Please enter your elevation: meters |
| Please indicate the seasons during which your site reports data: |
| Winter: ☐ Spring: ☐ Summer: ☐ Fall: ☐ |

Note that some of the text fields are blank because no default value attribute has been specified. These require user input, and there is no way to establish ahead of time what this input might be. However, it may still be worthwhile in some cases to provide a default value if that might help the user to understand what is required. When the allowed input choices can be limited ahead of time by the creator of the document, it is appropriate to use radio buttons and checkboxes. You can create as many different combinations of these kinds of field as your application needs.

Each group of radio and checkbox buttons has its own unique field name and, within each group, each button should have its own value. In Document 3.3, there are two radio button groups, named NS and EW. It is important to specify a value for each button, because the value of the checked button will be captured when the contents of the form are submitted to a recipient's e-mail address. This is demonstrated in the modified version of this document in Section 3.5. Default values for the radio field can be specified by using the checked attribute.

When you access the document, the button with the `checked` attribute will be "on." You can change it by clicking on another of the buttons in the group.

The same basic rules apply to `checkbox` fields. You can have more than one group of checkboxes, each with its unique name. The only difference is that you can select as many boxes as you like within each group.

3.3 Creating Pull-Down Lists

A common feature on Web pages that use forms is a pull-down list, which provides another way to limit the input choices a user can make on a form. The implementation described here is similar to a group of radio buttons in the sense that only one item can be selected from a list. This can simplify a document interface and eliminate the need for some input checking that might otherwise have to be done if a user is free to type whatever he/she likes in an input field. For example, creating a pull-down list of the months of the year eliminates the need for a user to type (and perhaps to mistype) the name of a month, as shown in Document 3.4.

Document 3.4 (`select.htm`)

```html
<html>
<head>
<title>Pull-Down List</title>
</head>
<body>
Select a month from this menu:
  <select name="testing">
    <option value="1" selected>January</option>
    <option value="2">February</option>
    <option value="3">March</option>
    <option value="4">April</option>
    <option value="5">May</option>
    <option value="6">June</option>
    <option value="7">July</option>
    <option value="8">August</option>
    <option value="9">September</option>
    <option value="10">October</option>
    <option value="11">November</option>
    <option value="12">December</option>
  </select>
</body>
</html>
```

In the output shown, the user has chosen the month of April, which is now high-lighted. The values of the value attribute can be, but do not have to be, the same as the text displayed for each option. In this case, the month values are numbers between 1 and 12, rather than the names of the months. Assigning the selected attribute to the first option means that "January" will be highlighted when the pull-down box is first displayed. For longer lists, the default format is for HTML to include a scroll bar alongside the list.

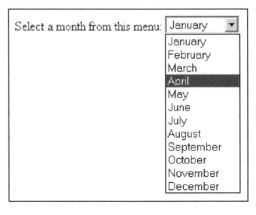

Although it is easy to create pull-down lists as well as groups of radio buttons and checkboxes, as described in Section 3.3, how a document will make use of the selections made is not obvious. However, as is shown in Chapter 4, JavaScript provides the required capabilities.

3.4 Combining Tables and Forms

In terms of organizing an interactive Web page, it is often helpful to create one or more tables in which the cell contents are fields in a form. Document 3.5 gives an example.

Document 3.5 (siteDefinition.htm)

```
<html>
<head>
<title>Observation Site Descriptions</title>
</head>
<body>
<form>
<table border="2" cellpadding="5" cellspacing="2"
  align="center">
  <caption><font size="+2">Observation Site
  Descriptions</font></caption>
  <tr bgcolor="lightblue">
```

```
  <th>Site #</th><th>Site Name</th><th>Latitude</th>
  <th>Longitude</td><th>Elevation</th>
</tr>
<tr bgcolor="palegreen">
  <td>Site 1</td>
  <td><input type="text" name="Name1" size="10"
    maxlength="10" value="Name1" /></td>
  <td><input type="text" name="Latitude1" size="10"
    maxlength="10"
    value="Latitude1" /></td>
  <td><input type="text" name="Longitude1" size="10"
    maxlength="10" value="Longitude1" /></td>
  <td><input type="text" name="Elevation1" size="10"
    maxlength="10" value="Elevation1" /></td>
</tr>
<tr bgcolor="ivory">
  <td>Site 2</td>
  <td><input type="text" name="Name2" size="10"
    maxlength="10" value="Name2" /></td>
  <td><input type="text" name="Latitude2" size="10"
    maxlength="10" value="Latitude2" /></td>
  <td><input type="text" name="Longitude2" size="10"
    maxlength="10" value="Longitude2" /></td>
  <td><input type="text" name="Elevation2" size="10"
    maxlength="10" value="Elevation2" /></td>
</tr>
<tr bgcolor="palegreen">
  <td>Site 3</td>
  <td><input type="text" name="Name3" size="10"
    maxlength="10" value="Name3" /></td>
  <td><input type="text" name="Latitude3" size="10"
    maxlength="10" value="Latitude3" /></td>
  <td><input type="text" name="Longitude3" size="10"
    maxlength="10" value="Longitude3" /></td>
  <td><input type="text" name="Elevation3" size="10"
    maxlength="10" value="Elevation3" /></td>
</tr>
<tr bgcolor="ivory">
  <td>Site 4</td>
  <td><input type="text" name="Name4" size="10"
    maxlength="10" value="Name4" /></td>
  <td><input type="text" name="Latitude4" size="10"
    maxlength="10" value="Latitude4" /></td>
  <td><input type="text" name="Longitude4" size="10"
    maxlength="10" value="Longitude4" /></td>
  <td><input type="text" name="Elevation4" size="10"
```

```
        maxlength="10" value="Elevation4" /></td>
  </tr>
  <tr bgcolor="palegreen">
    <td>Site 5</td>
    <td><input type="text" name="Name5" size="10"
        maxlength="10" value="Name5" /></td>
    <td><input type="text" name="Latitude5" size="10"
        maxlength="10" value="Latitude5" /></td>
    <td><input type="text" name="Longitude5" size="10"
        maxlength="10" value="Longitude5" /></td>
    <td><input type="text" name="Elevation5" size="10"
        maxlength="10" value="Elevation5" /></td>
  </tr>
</table>
</form>
</body>
</html>
```

Observation Site Descriptions

Site #	Site Name	Latitude	Longitude	Elevation
Site 1	Name1	Latitude1	Longitude1	Elevation1
Site 2	Name2	Latitude2	Longitude2	Elevation2
Site 3	Name3	Latitude3	Longitude3	Elevation3
Site 4	Name4	Latitude4	Longitude4	Elevation4
Site 5	Name5	Latitude5	Longitude5	Elevation5

The output is shown with the original default field names, before a user starts to add new values.

Although it may seem like a lot of work to create Document 3.5, the task is greatly simplified by copying and pasting information for the rows. When you access this page, the Tab key moves from field to field but skips the first column, which is just fixed text. The user of the page can change the default values of all the input text boxes.

3.5 E-Mailing the Contents of Forms

Document 3.3 would be much more useful if the location information provided by the user could be sent to the creator of the document. In general, if the basic purpose of forms is to provide an interactive interface between the user of a Web page and its creator, there has to be a way to transmit the user-supplied information on a form back to the creator. Remember that HTML/JavaScript constitutes a purely client-side environment. However, it is possible to use the `form action="mailto..."` and `method` attributes to send the contents of a form indirectly to the originator of the form (or some other specified destination) by using the client computer's e-mail utility.

In principle, this is easy to do, but the method described here is not very reliable. It may be necessary first to resolve conflicts between a user's browser and e-mail utility that have nothing to do with the contents of the Web page itself, or it may simply not be possible to get this method to work across some networks and platforms.

The following is the way to direct the contents of a form to a specified e-mail address, at least in principle!

```
<form method="post"
  action="mailto:my_mail@myuniversity.edu"
  enctype="text/plain">
```

Document 3.6 is a modification of Document 3.3 that allows a user to e-mail the contents of the form to a specified address.

Document 3.6 (`location2.htm`)

```
<html>
<head>
<title>Location information</title>
</head>
<body bgcolor="ivory">
<form method="post"
    action="mailto:my_mail@university.edu"
    enctype="text/plain">
      Please enter your last name:
    <input type="text" name="last_name" size="20"
      maxlength="20" /><br/>
      Please enter your latitude:
      <input type="text" name="lat" size="7"
       maxlength="7" />
       N <input type="radio" name="NS" value="N" />
```

```
   or S <input type="radio" name="NS" value="S" /><br/>
       Please enter your longitude:
       <input type="text" name="lon" size="8"
         maxlength="8" />
        E <input type="radio" name="EW" value="E">
   or W <input type="radio" name="EW" value="W" /><br/>
       Please enter your elevation:
       <input type="text" name="elevation" size="8"
         maxlength="8" /> meters<br/>
       <input type="submit"
         value="Click here to send your data." />
</form>
</body>
</html>
```

Please enter your last name: Brooks

Please enter your latitude: 40   N ⊙ or S ○

Please enter your longitude: 75   E ○ or W ⊙

Please enter your elevation: 15   meters

Click here to send your data.

After entering all values, the user clicks on the labeled submit button, and the contents of the form *should* be sent to the specified e-mail address. In order to try this document, you must install it on a Web page where it can be accessed online. (It will not work if you try to submit the form locally from an HTML editor, for example.) Sometimes, the submit button may not *actually* work. When you click on the submit button, it may *appear* that the data have been sent, but the e-mail never arrives. When this happens, the problem lies not with the document, but with the relationship between your browser and your e-mail utility. In some cases, it may not be possible to submit forms in this way from your computer.

When the form has been submitted successfully, the field names and values arrive in the body of an e-mail message. The example shown in the screen rendering produces this result:

```
last_name=Brooks
lat=40
NS=N
lon=75
```

```
EW=W
elevation=15
```

The names are the field names given in the document and the values are, of course, the values entered by the user.

3.6 The List Elements

As shown above, the `table` and `form` elements are used as tools for organizing Web pages. Elements for creating lists provide another way to impose formatting on related content. Table 3.2 gives a brief summary of three kinds of lists.

Table 3.2. HTML list elements

Description	HTML Tags	Use
Definition (or glossary)	`<dl> ... </dl>`	For a list that includes names and extensive descriptions
Ordered	`<ol> ... </ol>`	When a list of things has to be numbered
Unordered	`<ul> ... </ul>`	For a list of "bulleted" items

For ordered and unordered lists, the `li` element is used to define items within the list. For definition lists (also called glossary lists), the `dt` element is used for the "name" and `dd` is used for the "definition." Document 3.7 shows how to use these list tags.

Document 3.7 (`lists.htm`)

```html
<html>
<head>
  <title>Using HTML Lists</title>
</head>
<body>
This page demonstrates the use of unordered, ordered, and
definition lists.
<ul>
  <li> Use unordered lists for "bulleted" items.</li>
  <li> Use ordered lists for numbered items. </li>
<li> Use definition lists for lists of items to be defined.
</li>
</ul>
Here are three ways to organize content in an HTML document:
<ol>
  <li>Use a table. </li>
```

```
<li>Use a list. </li>
<li>Use <font face="courier">&lt;pre&gt; ...
&lt;/pre&gt;</font> tags.  </li>
</ol>
This is a way to produce a neatly formatted glossary list.
<dl>
  <dt><strong>definition list</strong>
    (<font face="courier">&lt;dl&gt;</font>)</dt>
  <dd>Use this to display a list of glossary items and their
definitions. </dd>
  <dt><strong>ordered list</strong>
    (<font face="courier">&lt;ol&gt;</font>) </dt>
  <dd>Use this to display a numbered list. </dd>
  <dt><strong>unordered list</strong>
    (<font face="courier">&lt;ul&gt;</font>)</dt>
  <dd>Use this to display a list of bulleted items. </dd>
</dl>
</body>
</html>
```

This page demonstrates the use of unordered, ordered, and definition lists.

- Use unordered lists for "bulleted" items.
- Use ordered lists for numbered items.
- Use definition lists for lists of items to be defined.

Here are three ways to organize content in an HTML document:

1. Use a table.
2. Use a list.
3. Use <pre> ... </pre> tags.

This is a way to produce a neatly formatted glossary list.

definition list (<dl>)
　　　　Use this to display a list of glossary items and their definitions.
ordered list (<ol>)
　　　　Use this to display a numbered list.
unordered list (<ul>)
　　　　Use this to display a list of bulleted items.

The use of these tags imposes a preset format for displaying list items. Blank lines are inserted before and after the list, with no <br /> or

<p> ... <p> tags required to separate the lists from other text in the document. For ordered and unordered lists, the list items themselves are indented. For the definition list, the items are not indented, but the "definitions" are indented. The contents of a list item can include text formatting elements. For example, in Document 3.7, the items in the definition list use the strong element to display the item name in a bold font. A list item can be an image, <img src="..." />, or a URL reference, <a href="...">.

Note the use of < and > to display the < and > characters in the document. (Recall that if you simply enter these characters, they will not be displayed on the screen because HTML will try to associate them with tags.)

There are some attributes associated with list elements that provide a little more control over the appearance of lists.

start="*n*"
Value: The integer *n* specifies the starting value of an ordered list. The default value is start="1".

type = "..."
Values: For unordered lists: "disc" (the default value), "square", "circle"
For ordered lists: "A" (uppercase letters), "a" (lowercase letters), "I" (uppercase Roman letters, "i" (lowercase Roman letters), "1" (numbers, the default value)

value = "*n*"
Value: The integer *n* specifies a numerical value for an item in an ordered list that overrides the default value. Subsequent list items will be renumbered starting at this value.

Finally, it is possible to combine list types to create more complicated list structures. Document 3.8 shows how list tags can be used to create the table of contents for a book.

Document 3.8 (bookContents.htm)

```
<html>
<title>Table of Contents for My Book</title>
<body>
<h2>Table of Contents for My Book</h2>
<ol>
<strong><li>Chapter One</strong></li>
```

```html
  <ol type="I">
    <li>Section 1.1</li>
      <ol type="i">
          <li>First Topic</li>
        <li>Second Topic</li>
          <ul type="circle">
            <li><em> subtopic 1</em></li>
            <li><em> subtopic 2</em></li>
          </ul>
      </ol>
    <li>Section 1.2</li>
    <li>Section 1.3</li>
  </ol>
<strong><li>Chapter Two</strong></li>
  <ol type="I">
    <li>Section 2.1</li>
  <ol type="i">
    <li>First Topic</li>
    <li>Second Topic</li>
      <ul type="circle">
        <li><em> subtopic 1</em></li>
        <li><em> subtopic 2</em></li>
      </ul>
  </ol>
    <li>Section 2.2</li>
    <li>Section 2.3</li>
  </ol>
<strong><li>Chapter Three</strong></li>
  <ol type="I">
    <li>Section 3.1</li>
      <ol type="i">
        <li>First Topic</li>
        <li>Second Topic</li>
          <ul type="circle">
            <li><em> subtopic 1</em></li>
            <li><em> subtopic 2</em></li>
            <li><em> subtopic 3</em></li>
          </ul>
      </ol>
    <li>Section 3.2</li>
    <li>Section 3.3</li>
      <ol type="i">
        <li>First Topic</li>
        <li>Second Topic</li>
      </ol>
    <li>Section 3.4</li>
  </ol>
</ol>
</body>
</html>
```

Note that if this list were used for an online book, for example, each list item could include a link to a URL or a hypertext link to another location within the same document.

Table of Contents for My Book

1. Chapter One
    I. Section 1.1
        i. First Topic
        ii. Second Topic
            ◇ *subtopic 1*
            ◇ *subtopic 2*
    II. Section 1.2
    III. Section 1.3
2. Chapter Two
    I. Section 2.1
        i. First Topic
        ii. Second Topic
            ◇ *subtopic 1*
            ◇ *subtopic 2*
    II. Section 2.2
    III. Section 2.3
3. Chapter Three
    I. Section 3.1
        i. First Topic
        ii. Second Topic
            ◇ *subtopic 1*
            ◇ *subtopic 2*
            ◇ *subtopic 3*
    II. Section 3.2
    III. Section 3.3
        i. First Topic
        ii. Second Topic
    IV. Section 3.4

3.7 Another Example

> Create a document that allows users to select observed cloud types from a list of possibilities. More than one cloud type can exist simultaneously. The categories are:
>
> High altitude: Cirrus, Cirrocumulus, Cirrostratus
> Mid altitude: Altostratus, Altocumulus
> Low altitude: Stratus, Stratocumulus, Cumulus
> Precipitation-producing: Nimbostratus, Cumulonimbus

A good way to organize this information is to use a table within a form. The form fields should be of type checkbox rather than radio because multiple selections are possible. Compare this problem with Document 3.2, in which a table was used to display just the cloud types.

Document 3.9 (cloud1.htm)

```
<html>
<head>
<title>Cloud Observations</title>
</head>
<body bgcolor="#aaddff">
<h1>Cloud Observations</h1>
<strong> Cloud Observations </strong>(Select as many cloud
types as observed.)
<br />
<form>
<table>
  <tr>
   <td><strong>High</strong> </td>
    <td>
     <input type="checkbox" name="high"
       value="Cirrus" /> Cirrus</td>
    <td>
     <input type="checkbox" name="high"
       value="Cirrocumulus" /> Cirrocumulus </td>
    <td>
      <input type="checkbox" name="high"
       value="Cirrostratus" /> Cirrostratus </td></tr>
  <tr>
    <td colspan="4"><hr noshade color="black" />
    </td></tr>
  <tr>
    <td> <strong>Middle</strong> </td>
    <td>
```

```
      <input type="checkbox" name="mid"
         value="Altostratus" /> Altostratus </td>
   <td>
      <input type="checkbox" name="mid"
         value="Altocumulus" /> Altocumulus</td></tr>
  <tr>
   <td colspan="4"><hr noshade color="black" />
    </td></tr>
  <tr>
   <td> <strong>Low</strong></td>
   <td>
   <input type="checkbox" name="low" value="Stratus" />
      Stratus</td>
   <td>
      <input type="checkbox" name="low"
        value="Stratocumulus" /> Stratocumulus</td>
   <td>
   <input type="checkbox" name="low" value="Cumulus" />
      Cumulus </td></tr>
  <tr>
   <td colspan="4"><hr noshade color="black" />
      </td></tr>
  <tr>
   <td> <strong>Rain-Producing </strong> </td>
   <td>
      <input type="checkbox" name="rain"
        value="Nimbostratus" /> Nimbostratus</td>
   <td>
      <input type="checkbox" name="rain"
        value="Cumulonimbus" /> Cumulonimbus </td></tr>
</table>

</form>
</body>
</html>
```

(See Color Example 6 for full-color output.)

In Document 3.9, checkboxes for the cloud types are organized into four groups, for high-, mid-, and low-altitude clouds, plus rain-producing clouds. Within each group, each checkbox has a name associated with it. As demonstrated in Chapter 5, this arrangement makes it possible for JavaScript to "poll" the checkboxes to see which clouds are observed within each group.

Note that the names given to each checkbox in Document 3.9 are the same as the text entered in the corresponding cell. This is only because these names and text are reasonable descriptions of the cell contents. In general, the text in the cell does not have to be the same as, or even related to, the value of the `name` attribute of the checkbox.

4. Fundamentals of the JavaScript Language

Chapter 4 presents the core programming capabilities of JavaScript. The topics include basic programming terminology and concepts, code structure, data and objects, variables, operators, mathematical and string-manipulation functions, decision-making structures, and constructs for repetitive calculations.

4.1 Capabilities of JavaScript

In the previous chapters I discussed the conceptual model through which a scripting language such as JavaScript interacts with an HTML document. In order to perform useful tasks within this environment, you must understand the capabilities and structure of JavaScript, as well as the programming fundamentals needed to apply these capabilities. Although an HTML document interface is still required to manage input and output, the material in this chapter reflects an attempt to minimize the details of interactions between JavaScript and HTML in favor of presenting the programming concepts.

JavaScript shares capabilities with other languages such as C/C++. In general, what are the capabilities of these kinds of languages? What kinds of tasks can programmers expect them to perform? These tasks are as follows:

1. Manage input and output

To be useful, any language must provide an input/output (I/O) interface with a user. When a computer program is executed or a script is interpreted (in the case of JavaScript, as a result of loading a Web page into a user's browser), the user provides input. The language instructs the user's computer to perform a task based on the input. The language then instructs the computer to display the results. A simple interface (for a text-based language such as C, for example) will accept keyboard input and display text output on a computer monitor. As noted several times throughout this book, JavaScript and HTML work together to provide an I/O interface.

2. Permit values to be manipulated in a symbolic way, independent of the way a particular computer stores that information internally

The entire thrust of high-level programming languages is to provide a symbolic interface between a computer and a user. This allows users to interact with a computer in a more natural way. Quantities can be given symbolic names and can then be accessed and manipulated through those names.

3. Perform arithmetic operations on numbers

A general-purpose language must enable a range of arithmetic operations on numbers. Although JavaScript is not intended as a "number-crunching" language for serious scientific computing, it does support many arithmetic operations and functions including, for example, trigonometric, logarithmic, and exponential functions. Thus, it is useful for a wide range of numerical calculations of interest in science and engineering.

4. Perform operations on characters and strings of characters

A great deal of the work JavaScript is asked to do involves characters and strings of characters rather than numbers. For example, JavaScript may be asked to compare a name provided as input against a predefined set of names. An HTML document is inherently character-based, so JavaScript must support the manipulation of characters and strings of characters, including interpreting the latter as numbers and vice versa. This is necessary because computers store numerical values in ways that differ fundamentally from the way characters are stored.

5. Make decisions based on comparing values

Computers cannot make decisions by "thinking" about multiple possibilities in a humanlike way. However, they can compare values and act on the results of those comparisons. Typically, programs compare values and then execute instructions based on the results of those comparisons. In particular, such decisions are often embedded in **branching structures** that execute one set of instructions to the exclusion of others, based on a comparison of values.

6. Perform repetitive calculations

Loop structures are used to allow computers to perform repetitive calculations. These calculations might be terminated after they have been executed a specified number of times, or they may be executed only until or while some set of conditions is satisfied.

4.2 Some Essential Terminology

The terminology of programming languages can be confusing for beginners. Nevertheless, it is essential to agree upon the meaning and use of terms in order to discuss programming concepts, especially because the programming-specific meaning of some terms must be distinguished from their everyday conversational use. Table 4.1 gives some essential terms and their definitions.

Table 4.1. Definitions of some essential programming language terms

Term	Definitions and Examples

Expression	A group of tokens and (usually) operators that can be evaluated as part of a statement to yield a value.
	`y + z`
	`"This is a string."`

Identifier	The name associated with a variable, object, or function. any allowed name, e.g., `x`, `getArea`, `my_name`, without embedded spaces
Keyword	A word that is part of a language and has a specific meaning. Keywords cannot be used as identifiers. `function`, `var`, `for`

Literal	A value (as opposed to an identifier) embedded in a script.
	`3.14159`
	`"Here's a string."`

Operator	A token that performs a mathematical or other operation. `=, +, -, *, /, %`
Program	Loosely, a series of statements or a compiled equivalent. In JavaScript, a "program" is better referred to as a script. Scripts are interpreted one line at a time, not compiled.
Reserved word	A word that might become part of a language. Reserved words should not be used as identifiers. `class`, `const`
Script	A series of statements written in JavaScript or some other scripting language.

Table 4.1. (Concluded.)

Term	Definitions and Examples
Statement	A command that changes the status of a program as it executes, by defining variables, changing the value of a variable, or modifying the order in which other statements are executed.
	`x = y + z;` `area=Math.PI*radius*radius;`
Token	An indivisible lexical unit defined within a programming language.
	all variables, keywords, operators, and literals
Variable	A place in memory that holds data and is represented by a unique identifier.
	(see "identifier")

Some of these terms identify the building blocks of a JavaScript script, starting with **tokens**:

tokens (identifiers, keywords, literals, operators) → expressions → statements → script

4.3 Structure of JavaScript Code

4.3.1 JavaScript Statements

Instructions in JavaScript are conveyed through a series of **statements** (usually) embedded in an HTML `<script>` ... `</script>` element. As indicated in the previous section, statements are built from expressions consisting of tokens. To begin a statement, simply start typing something that follows the syntax rules of JavaScript. In general, when it is time to terminate a programming language statement, there are two choices. One choice is to press the Enter or Return key on your keyboard. This will terminate both the physical line and the statement, which means that each physical line can contain no more than one statement. (It could be a blank line with no statement at all.) The second choice is to use a unique **terminating character** to mark the end of a statement.

As a matter of syntax, JavaScript allows both these choices. An "end of line" mark (created by pushing the Enter or Return key) will mark the end of a statement. Owing to JavaScript's roots in C/C++, the preferred syntax is to terminate each statement with a semicolon. As a matter of style, JavaScript statements in this book will *always* terminate with a semicolon. As a bonus, this style choice allows multiple statements to appear on the same line.

A set of JavaScript statements is called a script. Presumably, the goal of a script is to perform some useful task. Thus, the implication of calling something a "script" is that it contains all the instructions required to complete a specified task. As noted in Chapter 1, even the simplest text editor can be used to create a script. Although there are many software tools for creating HTML/JavaScript documents, they are all just conveniences and are never actually required.

JavaScript is a **free-format language**, which means that statements can appear anywhere on a line. As long as you terminate each statement with a semicolon, you can even put multiple statements on a single line. This flexibility is *supposed* to encourage the writing of code that is logically organized and easy to read. Good programmers always adopt a consistent approach to the layout of their code. Hopefully, the examples in this text will point the way to producing easily readable code.

4.3.2 Statement Blocks

Often, several code statements are grouped together in a **statement block**. These blocks begin and end with curly brackets:

```
{
    {statements go here}
}
```

Later in this chapter, we will see several examples of how to use statement blocks.

4.3.3 Comments

Comments are an essential part of good programming style, no matter what the language. Comments are inserted into code by surrounding them by certain characters that will always be interpreted unambiguously as marking the beginning or end of a comment. JavaScript supports two kinds of comments: single- and multiple-line. You can use either or both of these comment formats within the same script, but they cannot be mixed in the same comment. Moreover, you cannot have "nested" multiple-line comments:

```
// This is a single-line comment.
/* This
```

```
    is a
            multiple-line
                comment.
*/
/* This code
/* will generate a syntax error! */
*/
```

A JavaScript interpreter ignores comments when it executes statements, so comments can occur on separate lines or on the same line as a statement. Comments started with a double slash cannot be placed at the beginning of a statement because JavaScript has no way of knowing where the comment ends and the code begins. However, the following code will work because there is an (invisible) "return" character at the end of the line that is interpreted as the end of the comment:

```
// The gravitational constant is
var g=9.8; // m/s^2
```

This will not work

```
// The gravitational constant is var g=9.8; // m/s^2
```

but this will:

```
/* The gravitational constant is */ var g=9.8; //m/s^2
```

It is easy to overlook the importance of including comments in your code. Intelligently commented code is easier to understand, both for you when you return to it at a later date and for others who have to examine your code. If you do not develop the habit of including comments in all your code, you will eventually be sorry!

There is a potential issue with comments that is unique to the Web environment: JavaScript code is downloaded to a user's computer as part of a Web page. The longer the code, the longer the download time. Thus, an HTML document with heavily commented JavaScript code takes longer to download than the same page without comments. However, any (extremely small!) potential performance penalty that might be associated with using comments is completely overshadowed by the importance of learning how to write readable and understandable code. In any case, there is no excuse for not including comments for the kinds of scripts you will be creating during a course based on this book.

4.4 Data and Objects

In general, programming languages can work with different kinds of information. Each kind of information is associated with a **data type**; each data type is stored differently within the programming environment; and each is associated with a specific set of operations. For example, it is obvious that you can add two numbers, but it is less obvious what (if anything) it means to associate an addition operation (3.3 + 12.9) with character literals (A + c). In the latter case, A and c are not being used as symbolic names, but as the "literal values" of A and c.

A principle central to all high-level programming languages is that discrete units of information called **variables** can be associated with specific locations in computer memory. Variables serve as "containers" for data. A data container is established by giving it a symbolic name, called an **identifier**—a process called **data declaration**. Once identifiers have been established with meaningful names, you can write code to manipulate information symbolically by using the identifier names, thereby freeing you from having to think directly about where information is actually stored in your computer's memory. (As a practical matter, you cannot figure out exactly where this information is stored even if you think you need to know.) In addition, this symbolic approach makes it possible to write scripts that will work on any computer that supports JavaScript.

4.4.1 Data Declarations and Variables

A basic programming rule, no matter what the language, is that variables must be declared before they are used elsewhere in a program. Data declaration assigns an identifier (a variable name) to a data container and associates the identifier with a particular location in your computer's memory. The allocation of memory is handled by the programming environment (in this case, your browser and its JavaScript application) and is rarely of any interest to you as a programmer.

The data declaration process, whether explicit or implicit, is required to enable a programming environment to manage its memory resources and perform appropriate operations. In JavaScript, the keyword `var` is used to declare variables and their identifiers. Consider the following code:

```
var g;
g=9.8;
g="gravitational acceleration";
```

Unlike some other languages such as C and C++, a single keyword serves to declare *all* variables, regardless of their data type. In the above example, the `var` statement asks the JavaScript interpreter to set aside space for a variable named g. At the time of the declaration, it is not yet clear what kind of information the identifier g is going to represent.

JavaScript is a **weakly typed language**, which means that the programmer has a great deal of latitude in associating an identifier with data of a particular type. Consider the second and third lines in the above code fragment. The second line associates g with the numerical value 9.8. The third associates g with the string "`gravitational acceleration`" and replaces the previous value with the new one. These statements imply that the "container" associated with the identifier g can hold anything you want it to hold and that you can change your mind about the nature as well as the value of the information held in the container. The data declaration statement in JavaScript reserves the *name* of an identifier associated with a data container, but not the nature of its *contents*. To put it another way, JavaScript *infers* data type from the current contents of a variable container. If the nature of the contents of the container (not just the value) is changed, then the data type associated with that container will change as well. If you use spreadsheets such as Excel, you are already familiar with this kind of data typing. When you enter content in a spreadsheet cell, the spreadsheet imposes its own default typing for the content—as a number or text, for example. If you enter something different in the same cell, the spreadsheet reinterprets the contents accordingly.

Owing to weak typing, it is not even necessary to use a variable declaration statement in JavaScript. The statement

```
pi=3.14159;
```

without a previous `var pi;` statement is an implicit data declaration for the variable identifier `pi`. Although this is allowed in JavaScript, implied declarations are poor programming practice in any language and you should avoid them in your code.

4.4.2 Data Types

JavaScript supports three basic data types (**primitives**): numbers, strings, and Boolean values. JavaScript does not distinguish between integers and real numbers; that is, it does not provide separate data types for integers and real numbers. Rather, JavaScript stores *all* numbers in a **floating point** format, which provides what is, in general, an approximation of the actual value. In contrast, integers, in languages that support a separate data type, are stored as exact values in a binary format. This distinction can have significant consequences in some kinds of numerical calculations.

Some languages, such as C/C++, have a separate data type for representing individual characters, from which string representations are built. JavaScript works essentially the other way around, with a single character being represented as a string variable of length one.

Boolean data have one of two values, `true` or `false`. Boolean variables can be assigned one of these two values, as in:

```
var x=true,y=false;
```

Note that the words `true` and `false` are values, not "names" (or string literals, as defined in the next section), so they are not surrounded by quote marks.

4.4.3 Literals

Literals are actual numbers, character strings, or Boolean values embedded in code. In the statement `var pi=3.14159;`, `3.14159` is a number literal. In the statement `var name="David Brooks";`, `"David Brooks"` is a string literal. The advantage of using literals is that their value is self-evident.

In general, it is good programming style not to use the same literal value in many places in your code. For example, rather than using the literal `3.1416` whenever you need the value of π, you should assign a value to the quantity π by using a data declaration statement `var pi=3.1416;`. Now you can insert the value of π anywhere in your program just by referring to its identifier. Suppose you declare `var B = 3.195;` and use this variable name in several places in your code. If, later on, you decide you have to change the value of `B` to `3.196`, you can make this change just once, in the data declaration statement, and the change will be made automatically everywhere `B` is used.

4.4.4 Case Sensitivity

JavaScript is case-sensitive, which means that all reserved words and identifiers must be spelled exactly as they have been defined. For example, `return` is *not* the same as `Return`. JavaScript understands the former spelling as a keyword, but the latter spelling has no special meaning. If you define a variable named `radius`, you cannot later change that spelling to `Radius` or `RADIUS`. Owing to case-sensitivity, you *could* define three separate identifiers as `radius`, `Radius`, and `RADIUS`, but this is potentially confusing and should be avoided.

There are two reasons why it is especially important to be very careful when you spell names in JavaScript. First, JavaScript does not require an explicit data declaration statement for variable identifiers. Thus, you could write the declaration statement `var taxes,income,rate;` and then, later in your script, type `texas=income*rate;`. This misspelling of `taxes` as `texas` would be an obvious mistake on your part, but JavaScript will not see anything wrong with what you have done.

Second, remember that HTML is *not* case-sensitive. Since you will be using HTML and JavaScript together in the same document, it is easy to forget this distinction between the two. Be careful!

4.4.5 Objects and Methods for Input and Output

In plain language usage, an object is a thing—any kind of thing. An object has properties. Perhaps it is a ball—round, 6 cm in diameter, shiny, and red. Objects can do things. A ball can roll and bounce. In the world of programming, objects are also things that have properties and can do things. For example, there is a `Math` object in JavaScript that knows about mathematical constants (properties) and how to do certain kinds of mathematical calculations (see Section 4.6). In programming terminology, implementations of actions associated with an object are called methods. For example, you might define a method to describe how high a ball will bounce when you drop it onto a hard surface.

The reason I am introducing objects now is that in order to see how JavaScript works, we have to display the results of calculations done in response to user input. For now, the `document.write()` method of the `document` object, first introduced in Chapter 1, or `window.alert()`, a method of the `window` object will be used to display output. It is not necessary to include the `window` object name, so it is all right simply to write `alert()`. The purpose of using these methods is to avoid, for now, worrying about the interface between JavaScript and `input` fields in HTML forms. In later chapters, these methods will be used much less frequently.

For the same reason, to avoid interactions with an HTML document, the `window.prompt()`, or `prompt()` method will be used for input. Both `prompt()` and `alert()` will be used much less frequently after JavaScript and HTML forms are integrated, although they will remain useful for monitoring the performance of scripts.

Suppose you wish to ask the user of a script to provide the radius of a circle. The statement

```
var radius=prompt("Give the radius of a circle: ");
```

results in a message box being opened on the user's monitor. The "undefined" message that may appear in the input box means that the variable `radius` does not currently have a value assigned to it. When a value is typed in the input box, that

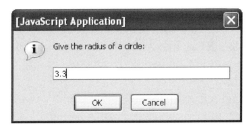

value will be assigned to the variable `radius`. Then, additional lines of code can be written to use that value. In subsequent sections of this chapter, I make frequent use of the `prompt()` method to get user input for a script. Document 4.1 shows how to use the `prompt()` method.

Document 4.1 (`circle.htm`)

```
<html>
<head>
<title>Calculate area of a circle.</title>
<script>
var radius=prompt("Give the radius of a circle: ");
radius=parseFloat(radius);
var area=Math.PI*radius*radius;
alert("The area of the circle with radius="+radius+" is
"+area+".");
</script>
</head>
<body>
</body>
</html>
```

[JavaScript Application]

⚠ The area of the circle with radius=3.3 is 34.21194399759285.

OK

Suppose you type 3.3 in the input box. The alert message box shown above will then appear on your screen.

The format of the `prompt()` and `alert()` windows is browser-dependent and cannot be changed from within your script.[1]

Note the shaded line in Document 4.1:

`radius=parseFloat(radius);`

---

[1] Some of my students complain that the alert box looks too much like a "warning," rather than an information window. For the examples in this chapter, you can use `document.write()` instead of `alert()` if that is your preference.

The purpose of parseFloat(), which is a "global" method not associated with a particular object, is to convert appropriate strings of characters into a numerical representation. (Global methods are discussed again in Chapter 6.) In Document 4.1, the variable radius is replaced by the output from the parseFloat() function. Why? Because anything entered in the prompt() input window is considered to be a string of characters, regardless of whether those characters "look" like a number. Often, scripts will work properly without the parseFloat() method, but there are many pitfalls, as will be discussed as appropriate in later examples. For now, suffice it to say that you should always apply parseFloat() to numerical data entered through a prompt(), regardless of whether it appears to be necessary.

4.4.6 String Methods

Owing to the importance of manipulating strings in interactions with HTML documents, JavaScript treats strings as objects and supports a long list of string-related methods. Table 4.2 lists some useful methods.

Table 4.2. Some useful methods for the String object

Method Name	Description and Examples
charAt(n)	Returns a string containing n^{th} character. n=0 returns leftmost character.
	"HTML".charAt(3); returns value of L.
charCodeAt(n)	Returns the base-10 ASCII code for the nth character. n=0 returns code for leftmost character.
	var s="acid",t; t=s.charCodeAt(0); T has value 97.
concat({two or more string arguments})	Concatenates (adds) the string arguments. (Equivalent to + operator with strings.)
	var s="I".concat(" love"," HTML."); S has value I love HTML.
fromCharCode(n₁[,n₂,,nₙ]) $n_1[,n_2,,n_n])$	Builds string from base-10 ASCII values.
	var s= String.fromCharCode(65,66,67); S has value ABC.

Table 4.2. (Concluded.)

Method Name	Description and Examples
indexOf(s[,n])	Returns index of first appearance of string s, at or after the optional starting index n. If n is not specified, search starts at index 0. Returns −1 if s is not found.
	`Excel.indexOf("x");` returns 1. `excel.indexOf("xce",2);` returns −1.
lastIndexOf(s[,n])	Returns index of last appearance of s, concluding the search at the optional second argument n. Returns −1 if s is not found.
	`excel.lastIndexOf("l");` returns a value of 4. `excel.lastIndexOf("e",2);` Returns a value of 0.
substr(m[,len])	Returns a new string containing a substring of the target string of length len, starting at index m. If len is not specified, the substring contains all characters from m to end of target string.
	`excel.substr(0,5);` returns excel. `excel.substr(2);` returns cel.
substring(m[,end])	Returns a new string containing a substring of the target string from index m up to but not including index end. If end is not specified, substring contains all characters from m to end of target string.
	`excel.substring(1,3);` returns ex.
toLowerCase()	Returns new string that converts all characters in target string to lower case.
	`var h="HTML";` `h=h.toLowerCase();` replaces h with the new value html.
toUpperCase()	Returns a new string that converts all characters in the target string to upper case.
	`var a="ascii",A;` `A=a.toUpperCase();` assigns a value of ASCII to A.

The methods of the `String` object in JavaScript can be applied directly to string primitives—variables or literals. Therefore, the reference to the `String` object name is rarely needed. This is in contrast to other objects, such as `Math`, which are discussed later in this chapter.

Character counting in strings, as for the `charAt(n)` method, starts from the left, at 0. Thus, `"HTML".charAt(3);` returns `"L"` and not `"M"`. There is only one string property: `length`. The value is set automatically whenever the contents of a string are changed; the value of `length` cannot otherwise be set. For example, `"ThisIsJavaScript".length;` returns a value of 16. It is important to understand that string methods do not change the contents of a string simply as a result of invoking that method. Rather, the method returns a value that must be assigned appropriately. Hence, in the example

```
var h="HTML";
h=h.toLowerCase();
```

the string variable h is replaced by a new value, `html`, through the assignment statement. In the example

```
var a="ascii",A;
A=a.toUpperCase();
```

the value of string variable a is unchanged, while the result of invoking the `toUpperCase()` method is assigned to the string variable A, which now has a value of `ASCII`.

4.5 Tokens, Operators, Expressions, and Statements

4.5.1 Tokens

As noted previously, tokens are the smallest lexical units of a language. One way to think about tokens is to consider how a script might be stored in compressed form. Each unique piece of information will be represented by a token; for example, variable name identifiers will be stored as tokens. The concept of tokens explains why `myname` or `my_name` are allowed variable names, but `my name` is not—`my name` will be interpreted as two separate names (two tokens).

4.5.2 Arithmetic Operators

Operators are also tokens. JavaScript operators, shown in Table 4.3, include arithmetic operators for addition, subtraction, multiplication, division, and the modulus operator for returning the remainder from division. These are all binary operators, which means that they require two operands, one to the left of the operator and one to the right. The addition and subtraction operators can also function as unary operators, with a single operand to the right of the operator; for example, $-x$.

With the exception of the modulus, or remainder, operator, these should all be familiar. The modulus operator works with either integer or real-number operands. (Remember that JavaScript does not support a separate integer data type.) The result of dividing 17 by 3 is 5 with a remainder of 2. The result of dividing 16.6 by 2.7 is 6 (6 times 2.7 = 16.2) with a remainder of $16.6 - 16.2 = 0.4$.

Table 4.3. JavaScript's arithmetic operators

Operator	Symbol	Examples	Precedence
Addition	+	3 + 4	2
Subtraction	–	z - 10	2
Multiplication	*	A*b	1
Division	/	z/3.333	1
Modulus (remainder)	%	17%3 (= 2), 16.6%2.7 (=0.4)	1

The addition operator also works as a concatenation operator for strings. The expression `var author = "David" + " " + "Brooks";` makes perfect sense to JavaScript and will give variable `author` the expected value of `"David Brooks"`. Note that the expression `"David" + "Brooks"` will produce the result `"DavidBrooks"`.

When JavaScript interprets an expression, it scans the expression from left to right one or more times. Operations implied by the presence of operators are evaluated according to **precedence rules**. Fortunately, these rules are the same ones that apply in algebraic expressions. Suppose $a = 3$, $b = 4$, and $c = 5$. What is the value of x in the algebraic expression $x = a + bc$? Based on precedence rules, multiplication and division operations are carried out before addition and subtraction. So, $x = 3 + 4 \cdot 5 = 3 + 20 = 23$. That is, a multiplication operation has precedence over an addition operation, so the addition operation is delayed until after the multiplication is performed, even though the addition operator is to the left of the multiplication operator. Parentheses are required to alter the precedence rules: $x = (3 + 4) \cdot 5 = 35$.

The same rules apply in JavaScript. As indicated in Table 4.3, multiplication and division (including the modulus operation) take precedence over addition and subtraction. Thus, in the following code,

```
var a=3,b=4,c=5;
var x,y;
x=a+b*c;
y=(a+b)*c;
```

the variable x has a value of 23. In the fourth statement, parentheses are used to override the natural order in which operations are evaluated, so y has a value of 35. The expression is evaluated from the innermost set of parentheses outward, so the a+b operation is performed before the multiplication by c.

4.5.3 The Assignment Operator

The JavaScript assignment operator is the symbol =. Thus, the JavaScript statement x=a+b; looks very much like the algebraic equation $x = a + b$. However, they are not at all the same thing! In programming, the assignment operator has a completely different meaning from the symbolic equality implied by the algebraic use of the = sign. In algebra, the equation $x = a + b$ defines a symbolic relationship among a, b, and x; whatever their values, x must be equal to the sum of a and b. Given values for a and b, you can determine the value of x. Given the values of x and a, you can solve for the value of b: $b = x - a$. Note also that $a + b = x$ is algebraically equivalent to $x = a + b$. In programming,

> **The meaning of the assignment operator is: "Evaluate the expression on the right side of the assignment operator and assign the result to the identifier on the left side of the assignment operator."**

For the statement x=a+b;, the specific meaning is "Assume that a and b have been given actual (often, but not always, numerical) values. Calculate their sum and assign the result to the identifier x."

With this definition of the assignment operator, it is clear that the JavaScript statement a+b=x; makes no sense, and will generate a syntax error. Why? Because:

> **Only an identifier can appear on the left side of the assignment operator.**

Finally, note that the algebraic expression $x = x + 1$ makes no sense at all because it is not possible for x to be equal itself plus 1. However, the JavaScript statement x=x+1; makes perfect sense. It means "Add 1 to the current value of x and then replace the value of x with this new value." Thus, as a result of executing these statements:

```
var x=5.5;
x=x+1;
```

x will have a value of 6.5.

It is sometimes difficult for beginning programmers to remember that an assignment statement is not the same thing as an algebraic equation. Although JavaScript allows you to represent data containers symbolically through the use of identifiers, the language does not understand the concepts of algebra. When JavaScript sees an assignment operator, all it knows how to do is evaluate the expression on the right side of the operator and assign that result to the identifier on the left side of the expression. In doing the expression evaluation, it assumes that every identifier has already been assigned an actual, and not just a symbolic, value.

As a result of how the assignment operator works, a general rule about assignment statements is:

> **An identifier should never appear on the right side of an assignment operator unless it has previously been assigned an appropriate value.**

Identifiers that do not follow this rule are called **uninitialized variables**. They are often given a value of 0 by default, but you should never violate the rule based on this assumption.

4.5.4 Shorthand Arithmetic/Assignment Operators

Table 4.4 shows some shorthand operators for combining arithmetic operations and assignments. They are popular among programmers because they are easy to write quickly, but their use is never actually required.

The increment operator (++) adds 1 to the value of the variable to which it is applied, and the decrement operator (--) subtracts 1. These operators are commonly used in looping structures, as discussed later in this chapter.

Table 4.4. Shorthand arithmetic/assignment operators

Operator	Implementation	Interpretation
+=	x+=y;	X=x+y;
-=	x-=y;	X=x-y;
*=	x*=y;	X=x*y;
/=	x/=y;	X=x/y;
%=	x%=y;	X=x%y;
++	x++; or ++x;	X=x+1;
--	y--; or --y;	X=x-1;

As shown in Table 4.4, you can apply these operators either before the variable name (pre-increment or pre-decrement) or after (post-increment or post-decrement). This choice can lead to some unexpected results. Consider Document 4.2.

Document 4.2 (incrementDecrement.htm)

```
<html>
<head>
<title>Increment/decrement operators</title>
<script>
      var x=3,y;
      y=(x++)+3;
      document.write("post-increment: y="+y+"<br />");
      document.write("x="+x+"<br />");
      x=3;
      y=(++x)+3;
      document.write("pre-increment: y="+y+"<br />");
      document.write("x="+x+"<br />");
</script>
</head>
<body>
</body>
</html>
```

```
post-increment: y=6
x=4
pre-increment: y=7
x=4
```

In the post-increment case, the value of x is incremented *after* the expression is evaluated to provide a value for y. In the pre-increment case, the value of x is incremented *before* the value of y is calculated. There would be a similar result for the decrement operator. For the most part, you should avoid combining the increment/decrement operators with other operations in a single expression. Furthermore, do not apply both pre- and post-operators at the same time (that is, do not write `++x++;` or `--x--;`) and do not apply these operators to the same variable more than once in an expression.

4.6 The JavaScript `Math` Object

In order for a programming language to be useful for scientific and engineering calculations, it has to have not only basic arithmetic operators, but also the ability to carry out other basic mathematics operations, such as you might find on a scientific calculator. In JavaScript, these operations are packaged as methods in the `Math` object, which also has properties that provide some useful mathematical values, such as π. The methods implement mathematical functions, such as trigonometric functions. With the single exception noted below, the methods have one or two real-number arguments and always return a real-number result, even when that result is a whole number that looks like an integer. Some methods and properties of the `Math` object are summarized in Table 4.5.

These methods must be used appropriately in order to produce meaningful results. For example, it makes no sense (at least in real-number mathematics) to ask `Math.sqrt()` to calculate the square root of a negative number. Fortunately or unfortunately, depending on your point of view, JavaScript is very forgiving about such abuses. It will return a "value" of `NaN` if you ask it to do an inappropriate calculation, but it will not describe the problem.

Trigonometric and inverse trigonometric functions always work in *radians*, not degrees. So `Math.sin(30);` will calculate the sine of 30 radians, not 30 degrees. This is an easy error to make, and it will not produce an error message because the requested calculation does not represent a problem from JavaScript's point of view. To convert from degrees to radians, multiply degrees by π/180.

When functions are called with very large or very small arguments, or when they should produce answers that are 0 (as in the sine of 0° or 180°) or approaching infinity (as in the tangent of 90°, problems can arise because of the imprecision inherent in real-number calculations. For example, `Math.sin(Math.PI);` will produce a value `1.2246e-16` rather than 0. (Try it and see.)

Table 4.5. Some properties and methods of the JavaScript `Math` object

Property	Description
`Math.E`	Base of the natural logarithm, e, 2.71828
`Math.LN2`	Natural logarithm of 2, 0.693147
`Math.LN10`	Natural logarithm of 10, 2.302585
`Math.LOG2E`	Log to the base 2 of e, 1.442695
`Math.LOG10E`	Log to the base 10 of e, 0.434294
`Math.PI`	π, 3.1415927
`Math.SQRT1_2`	Square root of ½, 0.7071067
`Math.SQRT2`	Square root of 2, 1.4142136
Method	**Returns**
`Math.abs(x)`	Absolute value of x
`Math.acos(x)`	Arc cosine of x, $\pm\pi$, for $-1 \leq x \leq 1$
`Math.asin(x)`	Arc sine of x, $\pm\pi/2$, for $-1 \leq x \leq 1$
`Math.atan(x)`	Arc tangent of x, $\pm\pi/2$, for $-\infty < x < \infty$ (compare with `Math.atan2(y,x)`)
`Math.atan2(y,x)`	Arc tangent of angle between x-axis and the point (x,y), measured counterclockwise (compare with `Math.atan(x)`)
`Math.ceil(x)`	Smallest integer greater than or equal to x
`Math.cos(x)`	Cosine of x, ± 1
`Math.exp(x)`	e to the x power (e^x)
`Math.floor(x)`	Greatest integer less than or equal to x
`Math.log(x)`	Natural (base e) logarithm of x, $x > 0$
`Math.max(x,y)`	Greater of x or y
`Math.min(x,y)`	Lesser of x or y
`Math.pow(x,y)`	x to the y power (x^y)
`Math.random()`	Random real number in the range [0,1]
`Math.round(x)`	x rounded to the nearest integer
`Math.sin(x)`	Sine of x
`Math.sqrt(x)`	Square root of x
`Math.tan(x)`	Tangent of x, $\pm\infty$

Despite the fact that "log" is often used to denote base 10 logarithms, with "ln" used for base e logarithms, the `Math.log()` object supports only natural (base e) logarithms and uses `log` rather than `ln`. Logarithms to some other base n can be calculated as

$$\log_n(x) = \log_e(x)/\log_e(n)$$

Base 10 logarithms are often used in engineering calculations. So, a JavaScript expression to calculate the base 10 logarithm of a variable x is

```
Math.log(x)/Math.log(10);
```

or, using the `Math.LN10` property,

```
Math.log(x)/Math.LN10;
```

The `Math` object methods mostly work just as you would expect. However, `random()` (the parentheses are required even though there is no calling argument) deserves a closer look. As is true for random number generators in all programming languages, JavaScript's `random()` method is really only a "pseudorandom" number generator. It relies on an algorithm that follows a pre-determined path whenever the method is used. The randomness results from "seeding" the algorithm with a starting value based on a value read from your computer system's internal clock. For all practical purposes, this "seed" value is not predictable, so it should allow generation of a sequence of numbers that appears to be random.

A call to an algorithm-driven random number generator such as `Math.random()` should generate a real number x randomly located within the interval $0 \le x < 1$. (That is, it is possible that x might be exactly 0, but not exactly 1.) This range can be expressed mathematically as $[0,1)$. Repeated calls to `Math.random()*n` should produce real numbers uniformly distributed over the interval $[0,n)$. However, practical applications of random numbers are more likely to require uniformly distributed integers over a specified range.

Caution is required when converting uniformly distributed real numbers to uniformly distributed integers. Some sources suggest

```
Math.round(Math.random*n+1) //Not a good idea!
```

This will produce integers in the range $[1,n]$, but those integers will *not* be uniformly distributed![2] One of the Chapter 4 exercises explores this problem

---

[2] Even *JavaScript: The Complete Reference*, the volume cited in Chapter 1, makes this mistake.

in more detail. See Document 4.3 for an appropriate approach to generating uniformly distributed integers.

Whenever a script contains many references to the Math object's properties and methods, it is convenient to use the with keyword. Within a with statement block, references to properties and methods do not have to be prefixed with the object name and dot operator.

```
with (Math) {
    {statements that refer to properties and/or methods of the Math
        object}
    var x=sin(.197);
}
```

Finally, it is interesting to note that you can create your own extensions to the Math object; for example, a method that correctly returns the value of an angle expressed in degrees rather than radians. These extensions exist only for the document in which they are defined, but you can save your own library of extensions, which can then be pasted into any script. For more information, see the exercises for Chapter 6.

Document 4.3 illustrates the use of some Math object methods. The for statement block is discussed later in this chapter. For now, its purpose should be clear from the output:

Document 4.3 (mathFunctions2.htm)

```
<html>
<head>
 <title>Demonstration of the Math object.</title>
<script language="javascript" type="text/javascript">
  for (var i=1; i<=10; i++)
    with (Math) {
     var x=floor(100*(random()%1))+1;
     document.write(x+" "+sqrt(x)+" "+pow(x,3)+"<br />");
    }
</script>
</head>
<body>
</body>
</html>
```

```
93 9.643650760992955 804357
73 8.54400374531753 389017
63 7.937253933193772 250047
69 8.306623862918074 328509
20 4.47213595499958 8000
95 9.746794344808963 857375
43 6.557438524302 79507
31 5.5677643628300215 29791
49 7 117649
10 3.1622776601683795 1000
```

This code will generate integer values of x in the range [1,100]. Why write Math.random()%1 rather than just Math.random()? If the random number generator happens to produce a value of exactly 1, the modulus operation replaces it with 0, because `1%1` equals 0. Any other number in the range [0,1) is unchanged by the modulus operation.[3]

The output from Document 4.3 illustrates an interesting point: Even though JavaScript does not have a data type for integers, it nonetheless knows how to display whole numbers not as real numbers with 0's to the right of a decimal point, but as integers. On the other hand, real numbers that are not whole numbers are typically displayed with 15 digits to the right of the decimal point! This is a consequence of how JavaScript stores numbers internally, but it is hardly ever desirable or meaningful to display this many digits.

Languages such as C/C++ have formatting options to gain more control over the appearance of output, but JavaScript provides only limited options. One solution makes use of the `Math.round()` method. If the statement (from Document 4.3)

```
document.write(x+" "+sqrt(x)+" "+pow(x,3)+"<br />");
```

is replaced with

```
document.write(x+" "+round(sqrt(x)*100)/100+" "+
  pow(x,3)+"<br />");
```

the output will be changed as shown, with only two digits to the right of the decimal point. Other values can be substituted for 100, as appropriate. The output is not simply truncated to the selected number of digits, but rounded appropriately, just as you would round numbers by hand. That is, if you wish to display the value of π with four digits to the right of the decimal point, both you and JavaScript would display 3.1415927 as 3.1416.

4	2	64
44	6.63	85184
75	8.66	421875
15	3.87	3375
38	6.16	54872
39	6.24	59319
18	4.24	5832
77	8.77	456533
57	7.55	185193
63	7.94	250047

A better solution makes use of the fact that JavaScript numbers are objects, with properties and methods. Some code that makes use of the `toFixed()` method for number objects would be as follows:

```
var x=2,n=3.3,z=3.777777;
document.write(x.toFixed(3)+"<br />");
document.write(n.toFixed(3)+"<br />");
document.write(z.toFixed(5)+"<br />");
```

---

[3] I have seen some online references claiming that some implementations of `Math.random()` might, in fact, occasionally produce a value exactly equal to 1.

```
/*
  This statement generates a syntax error.
  document.write(2.toFixed(3)+"<br />");
  but these work.
*/
document.write((7).toFixed(2)+"<br />");
document.write(13..toFixed(2)+"<br />");
```

The displayed results are:

```
2.000
3.300
3.77778
7.00
13.00
```

Note that you can use `toFixed()` to retain 0's to the right of the decimal point even for whole numbers, which you cannot do when you use `Math.round()`. Thus, `toFixed()` is probably the best way to exert some control over the appearance of JavaScript output.

4.7 Comparison Operators and Decision-Making Structures

4.7.1 Relational and Logical Operators

As noted at the beginning of this chapter, a programming language should be able to make decisions based on comparing values. JavaScript provides a set of operators for comparing values and a syntax for taking actions based on the results of comparisons. Table 4.6 summarizes JavaScript's **relational** and **logical operators**.

Some of these operators are familiar from mathematics. When two characters are required, it is because some mathematical symbols are not standard keyboard characters.

4.7.2 The `if` Construct (Branching Structures)

Branching structures are based on a translation into programming syntax of spoken-language statements such as: "If x is greater than y, then let $z = 10$, otherwise let $z = 0$" or "If today is Tuesday, I should be in class." Translating such statements into relational and logical tests makes it possible to build decision-making capabilities into a programming language.

Table 4.6. Relational and logical operators

Operator	Interpretation	Math Symbol	Precedence	Example	Value
Relational					
<	Less than	<	2	-3.3<0	true
>	Greater than	>	2	17.7>17.5	true
>=	Greater than or equal to	≥	2	17.7>=17.7	true
<=	Less than or equal to	≤	2	17.6<=17.7	true
==	Equal to, allowing for type conversion	=	3	9=="9"	true
===	Equal to, no type conversion	=	3	9==="9" "a"==="a"	false true
!=	Not equal to, allowing for type conversion	≠	3	9!="8" 9!="9"	true false
!==	Not equal to, no type conversion	≠	3	9!=="9"	true
Logical					
&&	AND		4	(x==3) && (y<0)	
||	OR		5	(x==3) || (z==4)	
!	NOT		1*	!(x==3)	

* Higher precedence than arithmetic operators.

JavaScript syntax is close to the spoken language, but of course it follows strict syntax rules. A generic outline would be as follows:

```
if  ({an expression. If true, statements are executed})
{
        {statements here}
```

```
}
// optionally
else if  ({an expression. If true, statements are executed})
{
        {statements here}
}
// optionally, more else if statements

// optionally
else
{
        {statements here}
}
```

The syntax requires only the if statement. The "then" word that you might use in conversation is implied—there is no then keyword in JavaScript. The expressions to be evaluated must be enclosed in parentheses. The else if's and else's are optional. The curly brackets are required to form a statement block whenever there is more than one statement for each branch.

If you consider an if structure as defining branches in a road that eventually rejoin at a main road, the minimum choice is a road with no branches, with the option to continue along the road toward your destination or to bypass the road completely.

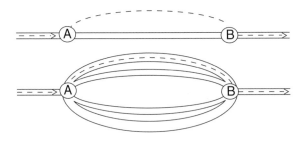

With multiple possible branches, it is important to understand that

Only the *first* branch of an if statement for which the expression evaluates as true will be taken.

To use the road analogy, once you select a road, you take only that road and no other.

This principle is illustrated in Document 4.4, which assigns a letter grade based on a 90/80/70/60 grading system. Suppose the numerical grade is 83. This is less than 90, so the first branch is not executed. However, 83 is greater than or equal to 80, so a letter grade of B is assigned. But 83 is also greater than or equal to 70. Does this mean that the letter grade is now reassigned to a C, etc.? No, because only the first true branch (assign a B) is executed; the subsequent branches are ignored.

Document 4.4 (grades.htm)

```
<html>
<head>
<title>Get letter grade</title>
<script language="javascript" type="text/javascript">
    var grade=
      parseFloat(prompt("What is your numerical grade?"));
  document.write("For a numerical grade of "+grade+
                 ", your letter grade is ");
  if (grade >= 90) document.write("A");
  else if (grade >= 80) document.write("B");
  else if (grade >= 70) document.write("C");
  else if (grade >= 60) document.write("D");
  else document.write("F");
  document.write(".");
</script>
</head>
<body>
</body>
</html>
```

[JavaScript Application]

ⓘ What is your numerical grade?

88

OK   Cancel

For a numerical grade of 88, your letter grade is B.

Note how identifier grade is given its value, with prompt() and parseFloat() combined in a single statement; for comparison, look again at Document 4.1. This script will actually work without applying parseFloat() because comparisons such as (grade >= 90) will apply an appropriate type conversion. However, neglecting to apply the parseFloat() requires JavaScript to compare "apples and oranges," and should be avoided both as a matter of good programming style and to prevent possible unforeseen problems in other circumstances.

Document 4.5 is another example of a calculation that uses an if structure. It calculates income tax when there are two tax rates, one of which applies to all income up to $50,000 and the other that applies to just that portion of income that is in excess of $50,000.

Document 4.5 (`taxes.htm`)

```
<html>
<head>
<title>Calculate income tax</title>
<script language="javascript" type="text/javascript">
var income=
prompt("Enter your income (no commas!): $");
income=parseFloat(income);
var tax,loRate=.17,hiRate=.37;
if (income<=50000.)
        tax=income*loRate;
else
        tax=50000.*loRate+(inco
me-50000.)*hiRate;
document.write("For an income
of $"+income+", your tax is
$"+tax.toFixed(2)+".");
</script>
</head>
</body>
</html>
```

[JavaScript Application]

(i) Enter your income (no commas!): $

73000

OK   Cancel

For an income of $73000, your tax is $17010.

For the example output, the tax is ($50,000)(0.17) + ($23,000)(0.37) = $17,010.00. The `toFixed(2)` method displays the result with two 0's to the right of the decimal point.

When comparisons get more complicated, you must be careful about how you form logical/relational expressions. Suppose you want your code to respond to the statement: "If today is Tuesday or Thursday, I should be in class." The proper implementation is:

```
if ((today == "Tuesday") || (today == "Thursday"))
```

If this expression is rewritten as

```
(today == "Tuesday" || "Thursday") // don't do it!
```

it has a value of `true` if `today` is `"Tuesday"` but a value of `"Thursday"` (rather than `false`) if `today` is `"Monday"`. This is not at all what you intended!

An alternate version of the original expression, without the two inner sets of parentheses, is

```
// poor style!
(today == "Tuesday" || today == "Thursday")
```

This will be interpreted correctly, but it depends on the fact that the equality operator has precedence over the OR operator. In cases like this, the use of "extra" parentheses, as in

```
((today == "Tuesday") || (today == "Thursday"))
```

is better programming style. It makes the order in which you wish the operations to be performed clear and also makes it unnecessary to memorize the precedence rules for relational and logical operators.

Finally, the expression

```
// don't do it!
(today = "Tuesday") || (today = "Thursday")
```

may *look* all right but, again, it is not at all what you intended because the equality operator has been replaced with an assignment operator. The expression has a value of "Thursday" rather than true.

> **Using an assignment operator (=) when you intend to use an equality operator (==) is a common programming mistake that is very hard to pinpoint because it does not generate a JavaScript error. Be careful!**

4.7.3 The switch Construct

There is one more type of branching construct that is useful for certain kinds of comparisons. Suppose we would like to write code that would tell a user how many days there are in a particular month.

Document 4.6 (daysInMonth.htm)

```
<html>
<head>
<title>Days in Month</title>
<script language="javascript" type="text/javascript">
var month=prompt("Give month (1-12): ");
switch (month) {
   case "1":
   case "3":
   case "5":
   case "7":
```

```
      case "8":
      case "10":
      case "12":
         alert("There are 31 days in this month."); break;
      case "4":
      case "6":
      case "9":
      case "11":
         alert("There are 30 days in this month."); break;
      case "2":
         alert("There are either 28 or 29 days in this
                  month."); break;
      default:
         alert("I do not understand your month entry.");
}
</script>
</head>
<body>
</body>
</html>
```

Although this code could be implemented with if syntax, the switch construct is perhaps a little more clear. The syntax should be clear from Document 4.6. The switch keyword is followed by an expression enclosed in parentheses. The possible values of the expression are enumerated in the case labels that follow. The "numbers" of the months are given as text because the value from prompt() is text. It will *not* work to replace the case statements with, for example, case 5: instead of case "5": because, unlike comparisons made with the == and other relational operators, no automatic type conversion will be performed. (See also the === and !== operators previously defined in Table 4.6.) If the line month=parseFloat(month); is inserted after the prompt, then the case values must all be numbers, and not text.

Each case and its value is followed by a colon. The values do not have to be in any particular order. The default keyword provides an opportunity to respond to unexpected or other values. The statements following the first case label whose value matches the expression are executed. Note that these statements are not enclosed in curly brackets. They are executed in order and will continue to execute subsequent statements that apply to other case values unless the break keyword appears as the last statement in a group of statements to be executed.

4.8 Loop Structures

The ability to perform repetitive calculations is important in computer algorithms and is enabled through the use of loop structures. Loops can be written to execute the same code statements a prescribed number of times, or they can be written so that loop execution (or termination) is based on conditions that change while statements in the loop are being executed. The former situation uses **count-controlled loops** and the latter uses **conditional loops**.

4.8.1 Count-Controlled Loops

Count-controlled loops are managed with the `for` keyword. The general syntax for such a loop is

```
for (counter= {expression giving on initial value of counter};
    {expression giving high (or low) value of counter};
    {expression controlling incrementing (or decrementing) of counter })
```

The `for` keyword is followed by three statements in parentheses. The first statement sets the initial value of a counter. You can give the identifier name—`counter` in the above example—using any name you like. The second expression sets conditions under which the loop should continue to execute, and the loop continues to execute as long as the value of the second expression is `true`. The third expression controls how the counter is incremented or decremented. It is up to you to make sure that these three related expressions are consistent and will actually cause the loop to terminate. For example, the loop

```
for (i=1; i=12; i+=2)
```

will never terminate because `i` will never equal 12. Perhaps you meant to write the second expression as `i<=12;`. If so, then the loop will execute for i=1, 3, 5, 7, 9, and 11.

Now, consider Document 4.7, which displays the integers 0–10, in order. The counter `k` is initialized to 1 and is incremented in steps of 1; the loop executes as long as `k` is less than 10. Use of the shortcut incrementing or decrementing operators, as in `k++`, is very common in `for` loops.

Document 4.7 (`counter2.htm`)

```
<html>
<head>
<title>Counter</title>
<script>
var k;
document.write("Here's a simple counter: "+"<br />");
for (k=0; k<=10; k++)
     document.write(k+"<br />");
</script>
</head>
<body>
</body>
</html>
```

```
Here's a simple counter:
0
1
2
3
4
5
6
7
8
9
10
```

For this example, a statement block enclosed in curly brackets following the `for` loop is not required because only one statement is executed in the loop. Document 4.8 shows a version of Document 4.6 that counts backward from 10.

Document 4.8 (`countdown2.htm`)

```
<html>
<head>
 <title>Countdown</title>
<script>
var k;
document.write("Start launch sequence!"
  +"<br />");
for (k=10; k>=0; k--)
  document.write(k+"<br />");
document.write("FIRE!!");
</script>
</head>
<body>
</body>
</html>
```

```
Start launch sequence!
10
9
8
7
6
5
4
3
2
1
0
FIRE!!
```

Recall that a `for` loop was used previously in Document 4.3. Now would be a good time to look back at that code and make sure you understand how that loop worked.

4.8.2 Conditional Loops

It is often the case that conditions under which repetitive calculations will or will not be executed cannot be determined in advance. Rather, conditions that control the execution or termination of a loop structure must be determined by values calculated inside the loop while the script is running. Such circumstances require conditional loops.

There are two kinds of conditional loops: **pre-test loops** and **post-test loops**. The statements in pre-test loops may or may not be executed at all, depending on the original values of loop-related variables. Post-test loops are always executed at least once, and the values of loop-related variables are tested at the end of the loop. The syntax is slightly different:

pre-test loop:

```
while  ({logical expression})  {
   {statements that result in changing the value of the pre-test logical
    expression}
}
```

post-test loop:

```
do  {
   {statements that result in changing the value of the post-test logical
    expression}
}  while  ({logical expression});
```

Conditional loops can be written either as post- or pre-test loops. The choice is based on how a problem is stated. Consider the following problem:

> A small elevator has a maximum capacity of 500 pounds. People waiting in line to enter the elevator are weighed. If they can get on the elevator without exceeding the load limit, they are allowed to enter. If not, the elevator leaves without trying to find someone who weighs less than the person currently first in line. If the elevator is overloaded, it crashes. It is possible that there might be a large gorilla in line, weighing more than 500 pounds. This gorilla should not be allowed on the elevator under any circumstances. Write a document that will supply random weights for people (or gorillas) waiting in line, control access to the elevator, and stop allowing people (or gorillas) to enter if the weight limit would be exceeded.

One solution to this problem is shown in Document 4.9.

Document 4.9 (`gorilla1.htm`)

```html
<html>
<head>
<title>The elevator problem (with gorillas).</title>
<script language="javascript" type="text/javascript">
  var totalWeight=0.,limitWeight=500.,maxWeight=550.;
  var newWeight;
  do {
    newWeight=Math.floor(Math.random()*(maxWeight+1));
    if ((totalWeight + newWeight) <= limitWeight) {
      totalWeight += newWeight;
      document.write(
        "New weight = " + newWeight + " total weight = "
        +totalWeight + "<br />");
      newWeight=0.;
    }
    else
      document.write("You weigh " + newWeight +
        " lb. I'm sorry, but you can't get on.");
  } while ((totalWeight + newWeight)
        <= limitWeight);
</script>
</head>
<body>
</body>
</html>
```

```
New weight = 191 total weight = 191
New weight = 154 total weight = 345
New weight = 151 total weight = 496
You weigh 108 lb. I'm sorry, but you can't get on.
```

This solution to the problem uses the `Math.random()` method to generate random weights between 0 and 500 pounds. The calculations are done inside a post-test loop. The code is arranged so that the effect of adding a new person to the elevator is tested before that person is allowed on the elevator. It is left as an end-of-chapter exercise to rewrite this as a pre-test loop.

In principle, count-controlled loops can also be written as conditional loops. (See the end-of-chapter exercises.) However, it is better programming style to reserve conditional loop structures for problems that actually need them. Clearly, Document 4.9 is such a problem because there is no way for the script to determine ahead of time what weights the `Math.random()` method will generate. Another example of a problem that demands a conditional loop calculation is Newton's algorithm for finding the square root of a number.

> Given a number n:
>
> 1. Make a guess (g) for the square root of n. $n/2$ is a reasonable guess.
> 2. Replace g with ($g + n/g$)/2.
> 3. Repeat step 2 until the absolute difference between g^2 and n is smaller than some specified value.

This algorithm is easy to write as a conditional loop. Consider Document 4.10:

Document 4.10 (newtonSqrt2.htm)

```html
<html>
<head>
<title>Newton's square root algorithm</title>
<script language="javascript" type="text/javascript">
var n=prompt("Enter a positive number:");
n=parseFloat(n);
var g=n/2;
do {
     g = (g + n/g)/2.;
} while (Math.abs(g*g-n) > 1e-5);
alert(g+" is the square root of "+n+".");
</script>
</head>
<body>
</body>
</html>
```

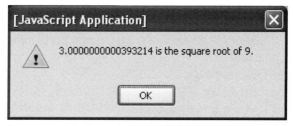

This algorithm is implemented as a post-test loop because a reasonable assumption is that the calculation inside the loop will always have to be done at least once. In fact, considering that the initial guess for the square root of n is $n/2$, this assumption is true for all values of n except 4. The statement `g=(g+n/g)/2;` is an excellent example of how an assignment operator differs from the same symbol (=) when it is used in an algebraic context. This kind of "replacement assignment" is often seen in conditional loops.

The terminating condition `while (Math.abs(g*g-n)>1e-5);` is important. It is not obvious whether g^2 will be larger or smaller than n. So, you must test the absolute value of $g^2 - n$ to ensure that the value being compared to 10^{-5} is always positive (because any negative number is less than $+10^{-5}$). This algorithm will work for any positive number. Note that the algorithm does not give *exactly* 3 as the square root of 9. On the other hand, if you calculate the square root of 4, it will give exactly 2. These kinds of discrepancies are a result of how numbers are stored and how numerical calculations are done. Newton's square root algorithm is a numerical approximation, so in general, it will *approach* the actual answer (within the specified accuracy), but will not necessarily give the exact answer for a perfect square. Except for annoying strings of zeros and digits—to the right of the 3 in the output shown here—these discrepancies are usually of no practical concern.

4.9 Using JavaScript to Change Values in Form Fields

In an interactive environment, you would like to be able to calculate new values based on user input. HTML form fields can serve both purposes: users can enter values and the document can use JavaScript to calculate new values for other fields. Consider the following problem:

> Atmospheric pressure decreases with elevation. When barometric pressure is given in weather reports, it is always referenced to sea level. (Otherwise it would not be possible to draw weather maps that show the movement of air masses.) Scientists often need to know the actual barometric pressure at a site, which is called station pressure. An approximate conversion from sea level pressure to station pressure is:
>
> $$P_{station} = P_{sea\ level} - h/9.2$$
>
> where pressure P is expressed in millibars and elevation h is expressed in meters.

Document 4.11 shows an HTML document that asks a user to provide elevation and sea level pressure and then calculates the station pressure. (Note that U.S. users will have to convert from inches of mercury to millibars.) This demonstrates several new HTML and JavaScript features, some of which are treated in more detail in subsequent chapters.

Document 4.11 (stationPressure.htm)

```
<html>
<head>
<title>Convert sea level pressure to station
pressure.</title>
<font size="+1">
<b>Convert sea level pressure to station pressure
(true pressure)</b></font><br /><br />
</head>
<body bgcolor="lightblue">
This application converts sea level pressure to
station pressure.<br />
Station pressure is the actual pressure at an
observer's observing site.<br />
It is always less than or equal to sea level pressure
(unless you are below<br />
sea level).
<br />
<form>
Fill in elevation and sea-level pressure:
<input type="text" name="elevation" value="0" size="8"
maxlength="7" /> (m)
<input type="text" name="sea_level_pressure"
value="1013.25" size="8" maxlength="7" /> (mbar) <br />
<input type="button" name="Calculate"
        value="Click here to get station pressure:"
onclick=
"result.value=
            parseFloat(sea_level_pressure.value)-
parseFloat(elevation.value)/9.2;" />
<input type="text" name="result" value="1013.25"
size="8" maxlength="7" /> (mbar)<br />
<input
type="reset"
value="Reset
all fields."
/>
</form>
</body>
</html>
```

Convert sea level pressure to station pressure (true pressure)

This application converts sea level pressure to station pressure.
Station pressure is the actual pressure at an observer's observing site.
It is always less than or equal to sea level pressure (unless you are below
sea level).

Fill in elevation and sea-level pressure: 0   (m) 1013.25   (mbar)

Click here to get station pressure: 1013.25   (mbar)

Reset all fields.

The HTML code in Document 4.11 provides default values for all the fields. The output reproduced here shows the default values, before a user has entered new values.

Earlier discussions noted that JavaScript script was often, but not always, contained within a `script` element in the `head` of a document. Based on Document 4.11, it is clear that JavaScript statements can appear elsewhere in a document, but it is not obvious that this should be so. For example, you could easily imagine a scenario in which JavaScript statements were allowed to exist *only* inside a `script` element.

The `"button"` field allows a user to initiate an action by clicking anywhere on the button. In this case, a click initiates the calculation of station pressure based on the values currently in the `elevation` and `sea_level_pressure` fields—either the default values or new values entered by the user. In order to respond to a user moving a mouse over the button field and clicking, HTML uses an **event handler**, an important means of providing interaction between a document and its user. Event handlers are attributes (of `input`) whose "values" consist of a set of JavaScript instructions enclosed in quotes. There are many event handlers, but in this chapter I use only `onclick`. (We will return to the topic of event handlers in Chapter 6.) In Document 4.11, the event to be "handled" is a click of a mouse when its cursor is somewhere in the screen space defined by the "Click here to get station pressure" button.

How is information transmitted from a form field to JavaScript? It will not work to use, for example, just the `elevation` name from

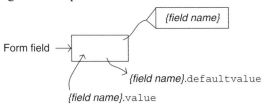

the form field. Why not? Because `elevation` is just the *name* of the field, not its *value*. Form fields have attributes, such as `name`, and those attributes have values, such as `elevation`. The attributes also have values, accessed through the "dot notation" shown. One of the values of a field name is its `defaultValue`, which is the value originally assigned to the form field in the HTML document (it could be blank).

Of interest here is `value`, the text entered in the form field. The value assigned to the `value` attribute contains the input for a calculation and will also receive calculated results in other form fields. Applying the `parseFloat(elevation.value)` method translates the text in `value` into a numerical value. Using just `elevation` as the argument for `parseFloat()` makes no sense at all from JavaScript's point of view. It

First JavaScript

This document was last modified on 11/16/2006 20:29:21

background = #90ee90
font = #ff00ff

Hello,world!

He said, "It's a beautiful day!"

Color Example 1. Document 1.4a, `HelloWorld3.htm`

This document was last modified on 11/16/2006 20:30:59

Our New House

Here's the status of our new house. (We know you're fascinated!)

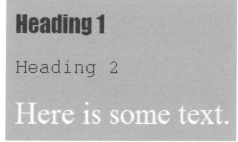

Color Example 2. Document 1.5, `house.htm`

Heading 1

Heading 2

Here is some text.

Color Example 3. Document 2.4b, `style2.htm`

This text should be blue on a red background.

This text should be red on a white background.

This text should be white on a blue background.

Color Example 4. Document 2.5b, `rwb.htm`

Results of radon testing

The table below shows some radon levels measured in residences. For values greater than or equal to 4 pCi/L, action should be taken to reduce the concentration of radon gas. For values greater than or equal to 3 pCi/L, retesting is recommended.

Location	Value, pCi/L	Comments
DB's house, basement	15.6	Action should be taken!
ID's house, 2nd floor bedroom	3.7	Should be retested.
FJ's house, 1st floor living room	0.9	No action required.
MB's house, 2nd floor bedroom	2.9	No action required.

Color Example 5. Document 3.1, `radonTable.htm`

Cloud Observations

Cloud Observations (Select as many cloud types as observed.)

High	☐ Cirrus	☐ Cirrocumulus	☐ Cirrostratus
Middle	☐ Altostratus	☐ Altocumulus	
Low	☐ Stratus	☐ Stratocumulus ☐ Cumulus	
Rain-Producing	☐ Nimbostratus	☐ Cumulonimbus	

Color Example 6. Document 3.9, `cloud1.htm`

may seem cumbersome to use this notation, but remember that the name assigned to an HTML form field is simply not the same thing as an identifier in JavaScript.

Once a mouse is clicked over the button field in Document 4.11, the JavaScript statement is executed. The application of `parseFloat()` to the values in the `elevation` and `sea_level_pressure` fields is required for the same reasons previously discussed for numerical values entered through `prompt()`. The distinction between text and numerical values is easy to forget because JavaScript often applies type conversions to text values on its own. In Document 4.11, the calculation for the `result` field *could* also be written as

```
result.value = sea_level_pressure.value -
               elevation.value/9.2; // Bad idea!
```

However, if you replace the "−" sign with a "+" sign, the numerical calculation will **not** be done! (Try it and see.) What is the difference? The "+" operator has a specific meaning when applied to strings (being interpreted as a concatenation operator), but the "−" operator does not. When it encounters a subtraction operator, JavaScript is "smart enough" to understand that the text values must be converted to numbers in order to carry out the specified action but, from the point of view of JavaScript, this is not necessary for the "addition" operator.

Type conversion issues also apply when results of a numerical operation are assigned to a form field name. Although `result.value=...` looks like an assignment of one numerical value to another, the numerical result must actually be converted back to text before it can be assigned to `result.value`. You might think that some kind of "convert this number to text" method is required, and in some sense it is, but you do not have to specify this conversion in your script.

Finally, clicking anywhere on the "Reset all fields" button sets all inputs back to their original values. JavaScript does this by accessing the `defaultValue` assigned to each field.

4.10 Another Example

The following is a simple algebraic calculation that is easy to implement:

For the quadratic equation $ax^2 + bx + c = 0$,

find the real roots:

$$r_1 = [-b + (b^2 - 4ac)^{1/2}]/2a \quad r_2 = [-b - (b^2 - 4ac)^{1/2}]/2a$$

The "a" coefficient must not be 0. If the discriminant $b^2 - 4ac = 0$, there is only one root. If $b^2 - 4ac$ is less than 0, there are no real roots.

Document 4.12 (`quadratic.htm`)

```
<html>
<head>
<title>Solving the Quadratic Equation</title>
</head><body><form>
Enter coefficients for ax<sup>2</sup> + bx + c = 0:
<br />
a = <input type="text" value="1" name="a" />
  (must not be 0)<br />
b = <input type="text" value="2" name="b" /><br />
c = <input type="text" value="-8" name="c" /><br />
click for r1 = <input type="text" value="0" name="r1"
onclick="var A=parseFloat(a.value),B=parseFloat(b.value),
C=parseFloat(c.value);
r1.value=(-B+Math.sqrt(B*B-4.*A*C))/2./A; " /><br />
click for r2 = <input type="text"
value="0" name="r2" onclick="var
A=parseFloat(a.value),B=parseFloat(b.value),
C=parseFloat(c.value);
r2.value=(-B-Math.sqrt(B*B-4.*A*C))/2./A; " /><br />
</form></body></html>
```

Enter coefficients for $ax^2 + bx + c = 0$:

a = 2   (must not be 0)
b = -9
c = 3
r1 = 4.1374586088176874
r2 = 0.36254139118231254

This is a workable solution to the problem, but it is certainly not elegant or thorough. (It is the kind of application you might write for your own use, but you might not want to distribute it globally on the Web!) For example, no check is performed on the discriminant to see if it is nonnegative before the `Math.sqrt()` method is applied. However, if the discriminant is negative, then JavaScript will simply assign a value of NaN to the result, which can be interpreted as a message that there are no real roots.

5. Using Arrays in HTML/JavaScript

Chapter 5 presents an introduction to arrays. It explains how to define arrays in JavaScript, how to use them, and how to use arrays to interact with an HTML form.

5.1 Basic Array Properties

The concept of **arrays** is extremely important in programming, as it provides a way to organize, access, and manipulate related quantities. It is important to form a mental model of how arrays are implemented, as shown in the sketch. It may be helpful to think of a post office analogy. The post office has a name, equivalent to the name of an array. Within the post office are numbered mail boxes. The numbers on the boxes correspond to array "addresses," called **indices**. The contents of the boxes correspond to array **elements**. In many programming languages, including JavaScript, the numbering of array boxes always begins at 0 rather than 1.

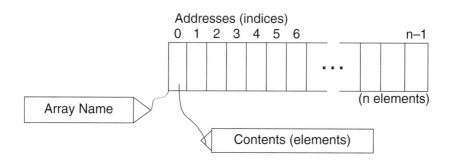

JavaScript supports an `Array` object for creating, using, and manipulating related values. The most important `Array` method is the

constructor new Array(), which allows us to create arrays. Syntax possibilities for doing this include:

```
var Array1 = new Array();
var Array2 = new Array(value_1,value_2,…,value_n);
var Array3 = new Array(10);
```

The first statement creates an empty array named Array1. The second statement creates an array of n elements with initial values as provided by the arguments. The third statement creates an array with 10 elements whose values are unspecified. In view of the fact that, as we will see, the declared size of an array is easily overridden, it makes little sense to declare an array using the syntax of the third statement.

It is not actually necessary to invoke the Array() constructor in order to create a JavaScript array. Each of these statements will create an array:

```
var Array1 = [];
var Array2 = [value_1,value_2,…,value_n];
var Array3 = [,,,,,,,,,];
```

Note the use of square brackets rather than parentheses in this syntax. The third statement, with nine commas, implies an empty array of 10 elements. The following syntax might be useful for declaring sparse (mostly empty) arrays:

```
var SparseArray = [1.1,,3.3,,,,];
```

Array elements can also be assigned by using variable names that already have appropriate values. The statements

```
var a=3.3, b=5.5, c=7.7;
var A = [a,b,c];
```

create array A with three elements equal to 3.3, 5.5, and 7.7.

"Square bracket" notation is also used to access the elements of an array. Individual array elements can appear to either the right or the left of an assignment operator, with the usual provision that those elements appearing on the right side of an assignment operator should already have been given appropriate values. That is, you can use assignment statements to assign values to undefined array elements or to change previously assigned values:

```
var a=3.3,b=5.5, c=7.7;
var A = [a,b,c];
var x,y=17.9,z=13.3;
x=A[0]+A[1]+A[2];
A[2]=y+z;
```

Array indices can be numbers, as in the above example, or identifiers, such as x[i], or even expressions, such as x[2*j+3], assuming the identifier or expression represents an integer value. If, for example, j = 2.5, the index is (2)(2.5) + 3 = 8, and this is an allowed index, assuming there are at least 9 elements in x. However, if j = 2.3, x[2*j+3] is undefined because (2)(2.3) + 3 is not a whole number. For the same reason, x[1] is defined, but x[4/3] is not.

Unlike some other languages, JavaScript allows an array declaration to be overridden later in the script. For the above example, it is easy to add another element:

```
A[3]=A[0]+A[1];
```

The current length of an array is contained in the length property. For the above example, A.length has a value of 4. The value of length is equal to the number of declared locations in the array, or to look at it another way, length gives the value of the next available array index. This is true regardless of whether any or all of the array elements have been assigned values. For example, the statements

```
var A = new Array(5);
alert(A.length);
```

display a value of 5 despite the fact that the array A is empty.

An interesting feature of JavaScript arrays is that not all elements must contain the same kind of data. Document 5.1 gives a simple example in which array elements are a mixture of numbers and text strings.

Document 5.1 (siteData.htm)

```
<html>
<head>
<title>Site Names</title>
```

```
<script>
  var siteID = ["Drexel",3,"home",101];
  var i;
  for (i=0; i<siteID.length; i++)
    document.write(i+",  "+siteID[i]+"<br />");
</script>
</head>
<body>
</body>
</html>
```

0, Drexel
1, 3
2, home
3, 101

Document 5.1 shows how the length property of an array is used to determine when the for loop should terminate. Remember that the index value of the last element in an array is always one less than the total number of elements in the array, which is why the terminating condition is i<siteID.length and not i<=siteID.length. The latter choice will not produce an error message, but it is an inappropriate choice for termination because the element A[A.length] does not exist.

As the number of elements in a JavaScript array can be expanded while a script is running, the code in Document 5.1 demonstrates the most reliable way to control a for loop when accessing arrays. Using the length property is always preferable to using a numeric literal as the terminating condition.

Note that it is also possible to use a for loop to access just parts of an array. For example,

```
for (i=1; i<A.length; i+=2) {
  . . .
}
```

accesses just the even elements of A – the 2nd, 4th, etc. (Starting the loop at an index of 1 first accesses the 2nd element of A.)

The code

```
for (i=A.length-1; i>=0; i--) {
  . . .
}
```

accesses the elements of A backward.

Another interesting feature of JavaScript is that you can use the assignment operator to assign one array name to another, but you have to be careful about how you interpret this action. Consider the following modification to Document 5.1:

```
var siteID = ["Drexel",3,"home",101];
var newSite = [];
```

```
var i;
newSite=siteID;
for (i=0; i<newSite.length; i++)
  alert(newSite[i]);
```

You could also have written var newSite = siteID;, which eliminates the need for the separate newSite=siteID; statement. A reasonable interpretation of such statements might be that newSite is an independent copy of siteID, stored in different memory locations from siteID. *However, this is **not** true!* The code does not actually create an independent copy of siteID. Instead, both siteID and newSite are now identified with the *same* data in memory because an array name does not literally represent the contents of the array. Rather, the name simply identifies a location in memory where the first element of the array is stored. If you assign one array name to another array name, all that happens is that the "box" in memory holding the array elements now has two "name tags" instead of just one. As a result, changes made to elements in either array will affect elements in the other array as well.

This interpretation of an array name also explains why the first element of an array is identified by an index of 0 rather than 1. The index is an offset— the "distance" from the memory location "pointed to" by the array name. For the very first element in an array, this offset is 0.

5.2 Some Operations on Arrays

There are some Array methods that are useful for the kinds of problems addressed in this book.

5.2.1 Manipulating Stacks and Queues

Stacks and **queues** are abstract data types— familiar to computer science students— that are used to store and retrieve data in a particular way. A stack uses a last-in first-out (LIFO) data storage model. You can think of it as a stack of dinner plates. You add dinner plates

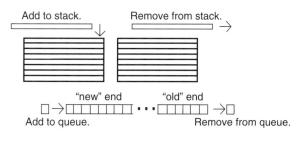

on the top of the stack, and when you retrieve one, you always take it from the top. As a result, the last value added on a stack is the first value retrieved.

A queue uses a first-in first-out (FIFO) data storage model, which operates like a queue (a line, in American English) of people waiting. A new person joins the line at the end, and people leave according to who has been in line the longest. Thus, a value removed from a queue is always the "oldest" value.

JavaScript arrays provide a very friendly environment for implementing stacks and queues because arrays can be resized dynamically while a script is running. However, the methods shown here for operating on stacks and queues may not work in all browsers. For example, they do not work in the internal browser supplied with the AceHTML freeware that I have used extensively for developing the code in this book.[1] You will just have to try the code to see if it works with your browser.

The push() and pop() methods are used for managing stacks and queues. push() adds ("pushes") the specified arguments to the end of the target array (the "top" of the stack), in order, as you would for either a stack or a queue. The length property is automatically updated. The pop() method (no calling arguments) removes ("pops") the last (most recent) element from the array, returns the value of that element, and decreases length by 1, as you would for a stack.

The shift() and unshift() methods are similar to push() and pop(), except that they operate from the front (index 0) of an array. shift() (no arguments) removes the first element from the array (as you would for a queue), returns the value of that element, shifts the remaining elements down one position, and decreases length by 1. The unshift() method shifts current array elements up one position for each argument, inserts its arguments in order at the beginning of the array, and increases length by 1 for each argument. This action would not be used with either a stack or queue. The use of these methods might seem backward because unshift() adds elements and shift() removes them.

To summarize:

• For a queue: use push() to add a new value at the end of the queue and shift() to remove the "oldest" value (the value at index 0).

---

[1] You can specify an external browser to use from within AceHTML, to replace its internal browser.

- For a stack: use `push()` to add a new value to the top of the stack and `pop()` to remove a value from the top of the stack.

        Documents 5.2 illustrates how to use these methods to treat an array first as a stack and then as a queue.

Document 5.2 (`stacksAndQueues.htm`)

```html
<html>
<head>
<title>Stacks and Queues</title>
<script language="javascript" type="text/javascript">
  var a=[1,3,5,7], i;
// Treat the array like a stack.
  document.write("STACK:" + a + " length of a = " +
a.length+"<br />");
  a.push(11,12,13);
  document.write(a + " length of a = " + a.length
  +"<br />");
  for (i=1; i<=3; i++) {
    a.pop();
    document.write(a + " length of a = " +
    a.length+"<br />");
  }
// Treat the array like a queue.
document.write("QUEUE:" + a + "    length of a = " +
a.length+"<br />");
  a.push(11,12,13);
  document.write(a + " length of a = " + a.length
  +"<br />");
  for (i=1; i<=3; i++) {
  a.shift();
  document.write(a + " length of a = " + a.length
  +"<br />");
  }
</script>
</head>
<body></body>
</html>
```

```
STACK:1,3,5,7 length of a = 4
1,3,5,7,11,12,13 length of a = 7
1,3,5,7,11,12 length of a = 6
1,3,5,7,11 length of a = 5
1,3,5,7 length of a = 4
QUEUE:1,3,5,7 length of a = 4
1,3,5,7,11,12,13 length of a = 7
3,5,7,11,12,13 length of a = 6
5,7,11,12,13 length of a = 5
7,11,12,13 length of a = 4
```

        Note the use of an entire array name in the `document.write()` parameter list. This automatically displays all the elements of the array, separated by commas.

5.2.2 Sorting

Sorting array elements in ascending or descending order is a fundamental computing task. However, it can be challenging to write efficient **sorting algorithms**. (Understanding and developing sorting algorithms is a standard topic in traditional programming courses.) Fortunately, JavaScript has an `Array` method, `sort()`, that will operate on arrays without much work on your part. Document 5.3 shows how to use this method to sort an array in ascending order. Unfortunately, as you will see, this code does not produce the expected result!

Document 5.3 (`sort.htm`)

```
<html>
<head>
<title>Sorting Arrays</title>
<script language="javascript" type="text/javascript">
var a=[7,5,13,3];
document.write(a + "    length of a = " + a.length+"<br />");
a.sort();
document.write(a + "    length of a = " + a.length+"<br />");
</script>
</head>
 <body>
</body>
</html>
```

```
7,5,13,3 length of a = 4
13,3,5,7 length of a = 4
```

Output from an `alert()` is displayed for and after application of the `sort()` method. However, the array is clearly not sorted, as 13 is not less than 3! It is apparent that the `sort()` method has performed a "lexical" sort based on the order of characters in the ASCII character sequence even when the characters represent numbers; the character "1" comes before the character "3" in this sequence in the same sense that "a" comes before "c" in the dictionary, and therefore, "ac" comes before "c." This result would be easier to understand if the values from the array came from a `prompt()` or from the `input` fields of a form, in which case it has already been demonstrated that "numbers" are treated like strings of characters. However, for sorting arrays of numbers, this result is clearly a disaster!

The `sort()` method can cause problems even with text. If, for example, you replace the array declaration with

```
var a=["zena","David","apple","pie"];
```

the result is still probably not what you intended. Uppercase letters come before lowercase

| zena.David.apple.pie length of a = 4 |
| David.apple.pie.zena length of a = 4 |

letters in the ASCII sequence, so "David" is still "less than" "apple."

The behavior of the `sort()` method constitutes a serious implementation problem. If you are sorting just text, you could consider using the `toUpperCase()` or `toLowerCase()` methods to convert all of the letters to either uppercase or lowercase letters prior to applying the `sort()` method, but this is not a very satisfying solution in general. A more comprehensive solution is to supply the `sort()` method with code for deciding whether one item is larger than, smaller than, or equal to another item in an array. This solution is addressed in Chapter 6.

5.3 Creating Two-Dimensional Arrays

Document 5.1 showed how to store the name (or ID) of a site (perhaps an observing site for collecting data) in an array. In this example, only a single value—the site identification—was stored in the array. It would be useful to have an array that would store multiple pieces of information about each site—perhaps its ID, longitude, latitude, and elevation. Table 5.1 gives some sample data.

Table 5.1. Site information to be stored in an array

Site ID	Latitude	Longitude	Elevation (m)
Drexel	39.955	−75.188	10
Home	40.178	−75.333	140
North Carolina	35.452	−81.022	246
Argentina	−34.617	−58.367	30
Netherlands	52.382	4.933	−1

A logical way to store these data is in some array equivalent of a table. One index could refer to the row and another to the column.

One way to do this, which is well-suited to JavaScript's capabilities, is to number the rows and to refer to the columns by name. The row numbering would then start at 0, followed by row 1 and the column Longitude would refer to the value −75.33 in Table 5.1.

Document 5.4 shows how to represent the data from Table 5.1 in an array with numbered rows and named columns.

Document 5.4 (siteData3.htm)

```
<html>
<head>
<title>"Multidimensional" arrays</title>
<script language="javascript" type="text/javascript">
var siteID = new Array();
function IDArray(ID,lat,lon,elev) {
   this.ID=ID;
   this.lat=lat;
   this.lon=lon;
   this.elev=elev;
}
siteID[0]=new IDArray("Drexel",39.955,-75.188,10.);
siteID[1]=new IDArray("home",40.178,-75.333,140.);
siteID[2]=new IDArray("NC",35.452,-81.022,246);
siteID[3]=new IDArray("Argentina",-34.617,-58.37,30.);
siteID[4]=new IDArray("Netherlands",52.382,4.933,-1);
var i;
for (i=0; i<siteID.length; i++) {
       document.write(siteID[i].ID+
       ", "+siteID[i].lat+", "+siteID[i].lon+",
"+siteID[i].elev+"<br />");
}
</script>
</head>
<body>
</body>
</html>
```

```
Drexel, 39.955, -75.188, 10
home, 40.178, -75.333, 140
NC, 35.452, -81.022, 246
Argentina, -34.617, -58.37, 30
Netherlands, 52.382, 4.933, -1
```

This code will make more sense after you read Chapter 6, which deals with JavaScript functions. Basically, each element of the array siteID is created as an object with properties, using the new keyword to reference an array constructor, function IDArray(). This function creates properties for the elements of siteID, with names that are appropriate for the values passed as arguments. It is convenient to use the same names both as "placeholders" for the arguments and for the property names themselves. However, this is *just* a convenience. Rewriting function IDArray() as

```
function IDArray(a,b,c,d) {
    this.ID=a; this.lat=b; this.lon=c; this.elev=d;
    ...
}
```

does not change the results. However, because the property names should be meaningful names that can easily be understood when they are used elsewhere in your script it makes sense to name the latitude property lat, and not something meaningless.

There is another, simpler, way to construct multidimensional arrays that makes more sense for certain kinds of problems:

> Define a 3 × 3 two-dimensional array of integers, with values 1–9, and display the contents row-by-row. The integer values should be arranged so they form a "magic square," defined as an n × n square matrix of integers, with values 1 through n^2, each of which appears once and only once, arranged so that each row and column and each main diagonal add to the same value. It can be shown that for a matrix of size $n \times n$, this value is $n(n^2 + 1)/2$. For a 3 × 3 matrix, the value is 15.

Some JavaScript code for constructing such a matrix, which can be addressed by row and column indices, would be as shown in Document 5.5:

Document 5.5 (magicSquare.htm)

```
<html>
<head>
<title>magic Square</title>
<script language="javascript" type="text/javascript">
        var a=[[8,1,6],[3,5,7],[4,9,2]];
        var r,c; //alert(a[0].length);
        for (r=0; r<a.length; r++) {
            for (c=0; c<a[0].length; c++)
                document.write(a[r][c]+" ");
            document.write("<br />");
        }
</script>
</head>
<body>
</body>
</html>
```

```
8 1 6
3 5 7
4 9 2
```

Note how the two-dimensional array is defined in the highlighted `var` statement. An array constructor could be used, but it is not necessary. Each element is itself defined as an array, using square bracket notation as shown. The number of rows is available from `a.length` and the number of columns from `a[0].length` (or with any other defined index in place of the 0). The number of rows can be different from the number of columns. There is no restriction on the content of arrays defined in this way; as is true for all JavaScript arrays, the elements can contain any combination of numbers and text. Access to the rows and columns uses indices within a pair of square brackets, within nested `for...` loops. The index identifiers `r` and `c` (for "row" and "column") make sense for this exercise, but they can be given any name you choose. In principle, it is possible to extend this code to higher-dimensional arrays, but the code will quickly become unwieldy!

It is left as a Chapter 6 exercise to write appropriate code for adding up the rows, columns, and diagonals of this square matrix to determine whether the integers form a magic square.

5.4. Using Arrays to Access the Contents of Forms

5.4.1 Accessing Values of `type="text"` Fields

Consider this generic problem: A form stores several columns of data in a table. You want the last row of the table to hold the sum of each of the columns. Based on previous material, you can give each form field a name: `r1c1`, `r2c1`, `r3c1`, etc. Then, you need to add each value:

```
parseFloat(r1c1.value)+parseFloat(r2c1.value)+ ...
```

This is not a very satisfying solution, if for no other reason than the fact that large tables require a *lot* of typing.

Fortunately, there is a more elegant alternative. When you create an HTML form, all the elements are automatically stored in an array called `elements`. You can access the contents of this array just as you would the contents of any other array. Consider the following very simple document:

Document 5.6 (`formArray.htm`)

```
<html>
<head>
<title>Using the elements[] array to access values in
forms.</title>
</head>
<body>
<form name="myform">
   A[0]<input type="text" value="3" /><br />
   A[1]<input type="text" value="2" /><br />
</form>
<script language="javascript" type="text/javascript">
for(var i=0; i<document.myform.elements.length; i++) {
   document.write("A["+i+"] =
"+document.myform.elements[i].value+"<br />");
}
</script>
</body>
</html>
```

```
A[0] 3
A[1] 2

A[0] = 3
A[1] = 2
```

First of all, note that these form fields have not been given names in the `<input />` tags. They *could* have names, but the point here is to avoid having to assign many different field names to items that can be treated as a unit, under a single name. Not surprisingly, the elements of the `elements` array are assigned, starting with index 0, in the order in which they appear in the form.

Previously, forms themselves were not given names. However, it is entirely possible that you might wish to have more than one group of form fields in a document, each of which would have its own `elements` array and could be accessed through its own name. Hence, the use of the `name` attribute in the `form` tag in Document 5.6. In this example, the use of "document" in, for example,

```
document.myform.elements[i].value;
```

is optional.

Document 5.7 shows another example of using the `elements[]` array to access form fields. With multiple columns, you will have to implement the `for` loop appropriately. For example, in a form that should be treated as two columns (assuming that those values are the first fields in the form), the index values 0, 2, 4, ... will access the left column and 1, 3, 5, ... will access the right column.

Document 5.7 (sumForm.htm)

```html
<html>
<head>
<title>Sum a column of values</title>
</head>
<body>
<form name="sumform">
  <input type="text" value="3.3" /><br />
  <input type="text" value="3.9" /><br />
  <input type="text" value="7.1" /><br />
Here is the sum of all the values.<br />
  <input type="text" name="sum" value="0"
    /><br />
</form>
<script language="javascript" type="text/javascript">
      var sum=0;
      for (var i=0;
i<(sumform.elements.length-1);
i++)
sum+=parseFloat(sumform.
elements[i].value);
sumform.elements[sumform.
elements.length-1].value=sum;
</script>
</body>
</html>
```

3.3
3.9
7.1
Here is the sum of all the values.
14.299999999999999

5.4.2 Accessing type="radio" and type="checkbox" Fields

Consider the following fragment from an HTML document:

```html
Employee is punctual:
  Y <input type="radio" name="punctual" value="Y"
    checked />     
  N <input type="radio" name="punctual" value="N"
    /><br />
```

This code defines a type="radio" field with two possible values. If you look at the elements array associated with the form containing this fragment, each field will be stored as a separate element in the elements array. However, what you really want to know is which button in the "punctual" group has been pressed. Similarly, with a group of type="checkbox" fields, you want to know which choices are selected. Conveniently, each group of radio buttons or checkboxes is associated with its own array. Document 5.8 provides some examples of how to use arrays to access the contents of radio buttons and checkboxes.

Document 5.8 (buttonAccess.htm)

```
<html>
<head>
<title>Accessing Radio Buttons and Checkboxes</title>
</head>
<body>
Access contents of form fields...<br />
<form>
Give name: <input type="text" name="Ename" size="15"
value="Mr. Bland" /><br />
Employee is punctual:
Y <input type="radio" name="punctual" value="Y"
    checked />     
N <input type="radio" name="punctual" value="N" /><br />
Employee likes these animals:
Dogs <input type="checkbox" name="animals" value="dogs" />
Cats <input type="checkbox" name="animals" value="cats"
    checked />
Boa constrictors <input type="checkbox" name="animals"
    value="boas" checked /><br />
<input type="button"
    value="Check here to examine form contents. "
  onclick="howMany.value=elements.length;
    contents.value=elements[parseFloat(n.value)].value;
      var i;
      if (punctual[0].checked)
          alert(Ename.value+' is always on time.');
      else
          alert(Ename.value+' is always late.');
      for (i=0; i<animals.length; i++) {
        if (animals[i].checked) alert(Ename.value+
          ' likes '+animals[i].value);
      };" /><br />
# elements: <input type="text" name="howMany"
    value="0" /><br />
Which one (0 to # elements - 1)? <input type="text" name="n"
    value="1" />
Contents: <input type="text" name="contents"
    value="--" /><br />
</form>
</body>
</html>
```

```
Access contents of form fields...

Give name: |Mr. Bland              |
Employee is punctual: Y  (•      N  ○
Employee likes these animals: Dogs □  Cats ☑  Boa constrictors ☑
         Check here to examine form contents.

# elements: |10                    |
Which one (0 to # elements - 1)? |0   |        Contents: |Mr. Bland     |
```

```
Microsoft Internet Explorer  [X]

    ⚠    Mr. Bland is always on time.

              [    OK    ]
```

The output shows the screen after the button box has been clicked and the first `alert()` box is displayed.

5.5 Hiding the Contents of a JavaScript Script

Basic security might seem to be the most obvious reason to hide part or all of a script. However, a better reason in the JavaScript context is to make it easy to modify or update part of a script without disturbing the HTML document of which it is a part; this is especially useful if the same script is used in several different HTML documents.

To do this, it is possible to save JavaScript code in a separate file that is referenced in an HTML document. Note that this does not overcome the limitation that a script is always loaded into a client computer when the HTML document containing the script is accessed. All that actually happens is that the "hidden" file is sent to the client computer and inserted into the script when the script is executed. Although this file is not visible when the HTML document source is viewed from a browser, it is certainly a mistake to assume that this provides any serious security protection for the hidden file.

Based on the discussion of arrays in the previous section, one obvious use for a hidden file is to hold data that will be used to build an array within a script. If these data are stored in a separate file, you can then keep them up to date by editing just the data file rather than an entire HTML document. Document 5.7 is a version of Document 5.4 in which the ID data are stored in a separate file.

Arrays are used to store values in memory and manipulate them while a program is running. With traditional programming languages, data are stored in a file that is "loaded" into memory to be read from and written to when needed. In the same way, a program can create new data to be stored permanently in a file that exists external to the program itself.

     However, this model does not work with HTML/JavaScript. Why not? Remember that a JavaScript script is loaded into a client computer when a Web page is accessed. The client computer has access only to the contents of this script. Hence, it is not possible to access data from a file that remains behind on the server computer. This limits the usefulness of JavaScript arrays for accessing large amounts of data stored in a central location. This restriction applies even when JavaScript is used locally on your own computer, because JavaScript simply does not provide the tools for accessing or creating external data files even when they reside physically on the same computer as the script that is running.

     The alternative is to send all the required data along as part of the script, which is a workable solution for small amounts of data that do not have to be protected in a secure environment. This solution works for both online and local applications of JavaScript. In a local environment, it is even reasonable to store large amounts of data, although there are some formatting issues for storing data. Unlike other languages, JavaScript cannot simply "read" data stored in a specified text format. Instead, as shown in Document 5.9, the data should be stored as part of an array definition.

Document 5.9 (siteData4.htm)

```
<html>
<head>
<title>"Multidimensional" arrays</title>
// This file defines the site characteristics.
<script language="javascript" src="site_data.dat">
</script>
<script language="javascript" type="text/javascript">
var i;
for (i=0; i<siteID.length; i++) {
      document.write(siteID[i].ID+
      ", "+siteID[i].lat+", "+siteID[i].lon+",
      "+siteID[i].elev+
      "<br />");
}
</script>
</head>
<body>
</body>
</html>
```

Data file `siteData.dat` for `siteData4.htm`:

```
var siteID = new Array();
function IDArray(ID,lat,lon,elev) {
  this.ID=ID;
  this.lat=lat;
  this.lon=lon;
  this.elev=elev;
}
siteID[0]=new IDArray("Drexel",39.955,-75.188,10.);
siteID[1]=new IDArray("home",40.178,-75.333,140.);
siteID[2]=new IDArray("NC",35.452,-81.022,246);
siteID[3]=new
  IDArray("Argentina",-34.617,-58.367,30.);
siteID[4]=new IDArray("Netherlands",52.382,4.933,-1);
```

The file `site_data.dat` is referenced within its own `script` element:

```
<script language="javascript" src="site_data.dat">
</script>
```

It is more typical to give such a "hidden" file a `.js` (for JavaScript) extension, but it is not required. In this case, the `.dat` extension seemed to more accurately reflect the purpose of the file.

The `siteData.dat` file does not hold just the raw site ID information. Rather, it holds the information plus all the code required to define an array holding this information. Although not necessary, it seemed a convenient approach to minimize the number of separate `<script>` ... `</script>` elements required. As JavaScript arrays are expandable while a script is running, there is no restriction on how many new sites can be added to the file or, for that matter, on how many sites can be removed.

5.6 Another Example

The following is a typical problem that involves comparing the contents of a form field against a set of predetermined values:

> Provide a form that asks a user for a password. Check his/her entry against a list of passwords and provide an appropriate message depending on whether the password is valid or not. (It is not necessary to take any action other than printing an appropriate message.)

Document 5.10 provides a "solution" to this problem, but without at all addressing the issue of password security. In fact, I chose this example to serve as a reminder that ***there is no security associated with anything sent as part of a JavaScript script!*** So, this is just a demonstration of how to search through a list of items to see if a user-specified item is present, rather than an application you would want to use to safeguard information.

Document 5.10 (`password1.htm`)

```
<html>
<head>
<title>Check a password</title>
<script language="javascript" type="text/javascript">
var PWArray=new Array();
PWArray[0]="mypass";
PWArray[1]="yourpass";
</script>
</head>
<body>
<form>
Enter your password: <input type="password" name="PW"
value=" "
onchange="var found=false; result.value='not OK';
 for (var i=0; i<PWArray.length; i++)
   if (PW.value == PWArray[i]) {
      found=true;
      result.value='OK';
   } " /><br />
(Tab to or click on this box to check your password.)<br />
<input type="text" name="result"
value="Click to
check password. " />
</form>
</body>
</html>
```

Enter your password: ●●●●●●
(Tab to or click on this box to check your password.)
OK

6. JavaScript Functions

Chapter 6 introduces the important concepts of functions in programming and shows how to integrate documents, forms, JavaScript, and functions to create a complete HTML/JavaScript problem-solving environment.

6.1 The Purpose of Functions in Programming

Functions are defined as units of code that accept input, perform operations on that input, and return one or more results. The built-in JavaScript methods discussed in Chapter 4 are examples of functions. For example, the Math.sin() method accepts a single value as input—an angle expressed in radians—and returns the sine of that value. User-defined functions also accept input, often more than one value, and return a value. They are an important concept in any programming language. Three reasons to use functions are as follows:

1. Organizing solutions to computational problems

  A problem to be solved on a computer often consists of several related parts, in which output from one part is used as input to the next part. Functions provide a mechanism for creating a code structure that reflects the nature of this kind of problem. By organizing code into a series of self-contained modules, and by controlling the flow of information among these modules, the problem can be solved in a logical fashion, one part at a time. Basically, this is a matter of separating large problems into smaller and more manageable parts.

2. Creating reusable code

  Often, identical calculations must be done several times within a program, but with different values. Functions allow you to write code to perform the calculations just once, using variable names as "placeholders" that will represent actual values when the function is used. Once a function has been written and tested, it can be used in other programs as well, allowing you to create a library of useful calculations.

3. Sharing authorship of large programming projects
Large programming projects often involve more than one person. When a project is broken down into several smaller tasks, individual programmers can work independently and then collaborate to assemble the finished product. Without the separation of tasks made possible by functions, this kind of collaborative approach would not be practical.

In general, functions are "called" (or "invoked," in the same sense as previously described for object methods) by passing values from a calling program (or another function) to the function. The function executes some operations and then returns a result:

In addition to providing a mechanism for modularizing the solution to a problem, functions play an important role in program design. The syntax of function implementation forces a programmer to think carefully about a problem and its solution: "What information is required to complete this task? What information is provided when the task is completed? What steps are required to solve the problem? What information must be provided by the user of a program? Can the problem be divided into smaller related parts? How does each of the parts relate to the others? Are the specified inputs and outputs for each part consistent with the relationships among the parts?" Once these questions are answered, the structure of a program should be clear. Often, working out an appropriate function structure is the hardest part of solving a computational problem.

6.2 Defining JavaScript Functions

Functions are essential for JavaScript programming. In fact, a large portion of all JavaScript code is written as functions called from HTML documents. One of JavaScript's first applications was to use functions to check values entered in forms. Inappropriate values are flagged and a warning message is displayed. Forms can be used in conjunction with functions for many kinds of calculations, as is done throughout this chapter.

It is important to understand how information is provided to, and extracted from, a function. The basic model, applicable to JavaScript and many other languages, is that a function resides in an isolated subset of computer memory. Communications with the contents of this space are strictly controlled and limited to specific pathways.

Input to a JavaScript function is controlled through the function's **parameter list**. Output is controlled through a statement starting with the `return` keyword. The syntax for a generic function is as follows:

```
function doSomething(input1,input2,input3,...) {
  var local1,local2,local3,...;
  local1 =  {an expression using one or more inputs...};
  local2 =
    {an expression using one or more inputs and (optionally) local1...};
  local3 =
    {an expression using one or more inputs and (optionally) local1
      and local2...};
  {Do some calculations here with some combination of parameters
      and local variables...};
  return {a value};
}
```

The `function` keyword is required at the beginning of every function, and every function must have a name. The naming convention used in this generic two-word function name, `doSomething`, is typical (but not required) in JavaScript: the first word starts in lowercase and the second and subsequent words in uppercase. Spaces between parts of a function name are not allowed, but underlines are permitted. So, for example, you could name the function `do_something`, but not `do something` (because `do something` is interpreted as two tokens rather than one). As in all aspects of programming, it will be helpful in your own work to settle on a function-naming convention and use it consistently.

The parameter list contains the names of one or more input parameters, separated by commas and enclosed in parentheses. These names are placeholders for input values passed to the function when it is called. Rarely, a function will have no values in its parameter list, but parentheses would still be required.

All the code in a function constitutes a statement block, enclosed in left and right curly brackets. The opening bracket can appear either at the end of the `function...` line or on the next line. Your code will be more easily readable if you adopt a consistent style of indenting the body of the code, as shown in the example.

Within the function, one or more **local variables** can be defined in statements that begin with the `var` keyword. Local variables are not actually required for many calculations, but code may be clearer if the results of intermediate calculations are stored in separate variables. In any event, the required calculations are done using appropriate combinations of the input parameters and local variables. The general programming rule that a variable should never be used until it has first been assigned a value applies equally to local variables in functions. To put it another way, a local variable should never appear on the right-hand side of an assignment operator until it has first appeared on the left.

The result of calculations performed in a function is returned to the place from which the function was called by using the `return` keyword in a statement. Only one `return` statement can be executed in a function. (A function can have more than one `return` statement, perhaps in various possible branches of an `if...` construct, but only one of these can actually be executed.) The value to be returned can also be declared as a local variable:

```
function doSomething(input1,input2,input3,...) {
    var local1,local2,local3,...,outputName;
    local1 = {an expression using one or more inputs...};
    local2 = {an expression using one or more inputs and
              (optionally) local1...};
    local3 = {an expression using one or more inputs and (optionally)
              local 1 and local2...};
    outputName = {do something with some combination of parameters
                 and local variables...};
    return outputName;
}
```

Here is a diagram of the JavaScript function model. The box represents the computer memory set aside for the function. This space and the operations carried out within it are not visible to the rest of a script, including to other functions within that script. Access to the function's memory space is available along only two paths.

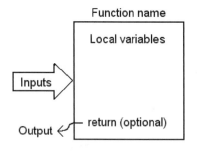

The large arrow represents the input pathway to the function, through its parameter list. The small arrow represents a single output from the function, generated as a result of a `return` statement. The two critical points are as follows:

> **The parameter list is a one-way path for input only. Information can be passed *in* to the function along this path, but no information passes *out* along this path.**

> **The `return` statement is a one-way path for a single value flowing *out* of the function.**

It is important to understand that the local variables defined within a function are invisible to the rest of your script, including to other functions. This means that you can select local variable names, assign values, and change those values without regard to what happens in other functions and elsewhere in a script, even when the same variable name is used elsewhere.

As is often the case, successful programming requires good mental pictures of how programming paradigms work. The function model shown here, including the restricted input/output paths and the protected nature of locally declared variables, is one of the most important paradigms in all of programming. It is what makes it possible to separate a large and complex computational problem into a series of smaller (and hopefully simpler) problems, linked through a series of function interfaces. This modularization makes even small scripts easier to write, and also makes it practical for large programming projects to be written, tested, and maintained by more than one person.

6.3 Using JavaScript Functions with HTML Forms

In a sense, all the previous material in this book has been directed toward this section. Why? Because the basic problem-solving model for the HTML/JavaScript environment is to use JavaScript functions together with forms in HTML documents.

The function model described in the preceding section would be very simple except for the fact that, in JavaScript, a value passed to a function through a parameter list can be one of three distinctly different things: a value (a character string or number), a form field, or an entire form. These are not interchangeable, and they all must be treated differently. In order to explain these differences, consider the simple problem of calculating the area of a circle. Given a radius r:

$$\text{area} = \pi r^2$$

Recall that `prompt()` and `alert()` and `document.write()` methods provided an I/O interface for these kinds of calculations in Chapter 4. Further on in that chapter, some JavaScript calculations were initiated as a result of using the `onclick` event handler in a `button` field. These approaches were acceptable at the time, but they are too limited to be good solutions for more complex problems. The following detailed analysis of several approaches to implementing this simple calculation in a function may seem tedious and unnecessary because the problem itself is so simple, but a thorough understanding of the analysis is absolutely essential to successful JavaScript programming.

6.3.1 Using Numerical Values as Input

A JavaScript function to solve the problem of calculating the area of a circle is

```
function getArea(r) {
        return Math.PI*r*r;
}
```

The parameter `r` is assumed to be a number representing the radius of a circle. The calculation is straightforward, using the `PI` property of the `Math` object (`Math.PI`). There is no exponential operator in JavaScript (r^2 cannot be represented as `r^2` as it could in a spreadsheet, for example), so `r` is just multiplied by itself.

It seems clear that you should be able to pass a value of the radius from a form field to `getArea()`. However, the examples in Chapter 4 provide ample evidence that caution is required! Consider this `input` element appearing within a form:

```
<form>
<input type="text" name="radius" maxlength ="6"
size="6" value="-99" />
. . .
```

Recall from Chapter 4 that information entered in form fields is always stored as text, even when the information is intended to be considered as numerical, and that the name of the field, `radius` in this case, is not the same as the value associated with this field. Thus, passing `radius` to `getArea` will *not* produce the desired result, nor will `radius.value`. Why not? Because `radius` is only the "value" of the `name` attribute, and `radius.value` is still only a character representation of the required numerical input.

You should not be surprised to learn that the calling argument to function `getArea()` must be `parseFloat(radius.value)`, as shown in Document 6.1.

Document 6.1 (`circle1.htm`)

```html
<html>
<head>
  <title>Circle Area (1)</title>
  <body bgcolor="#99ccff">
  <script language="javascript" type="text/javascript">
    function getArea(r) {
      return Math.PI*r*r;
    }
  </script>
</head>
<h3>Circle Area (1)</h3>
<p>
<form>
  Enter radius, then press tab key or click on "area"
    box.<br />
  radius (cm):
  <input type="text" name="radius" size="6" maxlength="7"
    value="-99",
    onblur="area.value=getArea(parseFloat(radius.value));"
 />
  area (cm<sup>2</sup>):
  <input type="text"  name="area" size="6" maxlength="7"
    value="-99" />
</form>
</body>
</html>
```

Circle Area (1)

Enter radius, then press tab key or click on "area" box.
radius (cm): 3   area (cm^2): 28.274333

The critical line of code in Document 6.1 is shaded because it initiates the call to `getArea()` through the `onblur` event handler, activated whenever the user of the form enters a value and then leaves the `radius` form field either by pressing the Tab key or by clicking elsewhere on the document. I summarize this and other event handlers later in this section.

The parameter in the call to `getArea()` is not just the field name `radius` but the `radius.value` property converted from a text string to a numerical value by the `parseFloat()` method. As in some examples shown in Chapter 4, the numerical result from the call to `getArea()` must be assigned to `area.value` and not just to `area`. If you try to do the latter, you will get a JavaScript error message. Again, `area.value` is actually a text string, not a number, but in order to display the result, JavaScript will automatically do this type conversion for you.

If you enter something in the `radius` field that cannot be interpreted as a number, then `radius.value` cannot be interpreted as a number. That means that the area of the circle cannot be calculated, and the `area` form field will display as NaN, for "not a number."

There is another subtlety worth noting about using functions with forms. Consider the following modification of Document 6.1:

```
<script language="javascript" type="text/javascript">
// UNACCEPTABLE CHOICE FOR FUNCTION NAME!
  function area(r) {
    return Math.PI*r*r;
  }
</script>
...
<form>
  Enter radius, then press tab key or click on "area"
    box.<br />
  radius (cm):
  <input type="text" name="radius" size="6"
    maxlength="7" value="-99",
    onblur =
"area.value=area(parseFloat(radius.value));" />
  area (cm<sup>2</sup>):
  <input type="text" name="area" size="6"
    maxlength="7" value="-99" />
...
```

In this code, the function name, `area`, is the same as a field name in the form. Although one could envision a programming environment in which this conflict could be resolved based on the context, this choice of names will produce a JavaScript error message and your code will *not* work. Therefore,

> **The names of functions should never be the same as the names of form `input` fields.**

The original Document 6.1 uses a typical style for naming functions: choose a prefix for the function name, as in `getArea()`, that would be an unlikely choice for an `input` field name.

6.3.2 Using Field Name value Attributes as Input

It is possible to apply the parseFloat() method inside a function, rather than in the call to the function. Consider the following modification of Document 6.1:

Document 6.2 (circle2.htm)

```
html>
<head>
<title>Circle Area (2)</title>
<body bgcolor="#99ccff">
<script language="javascript" type ="text/javascript">
      function getArea(r) {
         var radius=parseFloat(r);
         return Math.PI*radius*radius;
      }
</script>
</head>
<h3>Circle Area (1)</h3>
<form>
  radius (cm):
  <input type="text" name="radius" size="6"
    maxlength="7" value="-99",
    onblur = "area.value=getArea(radius.value);" />
  area (cm<sup>2</sup>):
  <input type="text"  name="area" size="6"
    maxlength="7" value="-99" />
</form>
</body>
```

6.3.3 Using Field Names as Input

It is also possible to pass an input attribute (a field name) to a function like getArea().Consider the modification of Document 6.2 shown below:

Document 6.3 (circle3.htm)

```
<html>
<head>
<title>Circle Area (3)</title>
<body bgcolor="#99ccff">
<script language="javascript" type ="text/javascript">
      function getArea(r) {
         var radius=parseFloat(r.value);
```

```
      return Math.PI*radius*radius;
    }
</script>
</head>
<h3>Circle Area (1)</h3>
<form>
  radius (cm):
  <input type="text" name="radius" size="6"
    maxlength="7" value="-99",
    onblur = "area.value=getArea(radius);" />
  area (cm<sup>2</sup>):
  <input type="text"  name="area" size="6"
    maxlength="7" value="-99" />
</form>
</body>
</html>
```

Note that the calling parameter to getArea() is the form field name radius. In the function's parameter list, this field name is r, and r now "points" in memory to the form field radius. The local variable name radius, defined in

```
var radius=parseFloat(r.value);
```

has a completely different meaning than it does in the HTML form. Remember that it is a local variable, which is "invisible" to the rest of the code; it is the translation into a numerical value of the text saved in the memory location pointed to by r. You may wish to avoid using identical names in this way, to minimize confusion, but it is done in Document 6.3 to make a specific point about how functions work.

Here, too, if r.value contains characters that cannot be interpreted as part of a number, the conversion cannot be done and a result of NaN will be returned.

6.3.4 Using Entire Forms as Input

There is a fourth way to write a function that calculates the area of a circle. Consider Document 6.4.

Document 6.4 (circle4.htm)

```
<html>
<head>
<title>Circle Area (4)</title>
<body bgcolor="#99ccff">
```

```
<script language="javascript" type ="text/javascript">
      function getArea(f) {
         var r=parseFloat(f.radius.value);
         f.area.value = Math.PI*r*r;
      }
</script>
</head>
<h1>Circle Area (3)</h1><p>
<form>
  Enter radius, then press tab key or click on "area"
box.<br />
   radius (cm):
   <input type="text" name="radius" size="6"
     maxlength="7" value="-99",
     onblur = "getArea(form);" />
   area (cm<sup>2</sup>):
   <input type="text"  name="area" size="6"
     maxlength="7" value="-99" />
</form>
</body>
</html>
```

In this version of getArea(), the entire form (actually, just information about where the form is located in computer memory) is passed to the function through the parameter name f. This name can be anything reasonable. There is no return statement. How, then, is the result of the calculation made available to the area form field? The answer lies in the following two statements:

```
var r=parseFloat(f.radius.value);
f.area.value = Math.PI*r*r;
```

The first statement extracts the numerical value of the radius, and the second modifies not the form parameter itself, but the value property of one of its fields (also converting the number back to text) Note that this approach requires that the function be aware of the names of the fields in the form passed to it as input, which is a major conceptual difference compared to the three previous approaches. The fact that the form and the JavaScript function are linked in this way is not a problem for self-contained documents such as the one being considered. The only disadvantage is that it may limit the use of the function in other scripts that may utilize different field names for the same physical quantities.

In the previous discussion of JavaScript's function model, it was clear that the parameter list acted as a one-way path for input to be passed to a function, but it could not be used to deliver output. Document 6.4 appears to violate this rule because the output has, in fact, been delivered back to the form "through" the parameter f. However, this result does not, in fact, compromise the model. When you pass a "value" to a function, you are actually passing memory addresses telling the function where particular parameters are stored. The function is allowed to make use of information stored in these addresses, but cannot use the addresses themselves. When the location of a form is passed as a parameter, what the function *can* do is modify the contents of fields stored in the form, which is what is done by the statement

```
f.area.value = Math.PI*r*r;
```

It is important to understand that the name f appearing in the function getArea(form) has *nothing* to do with names used in the HTML document. This is a consequence of the "protected" environment created by a function definition, in which names defined within the function are invisible to the rest of a document and script. In fact, it would be acceptable from the point of view of JavaScript to use form as a parameter name, although this might not be a good choice as a matter of style.

The ability of a function to modify fields within a form is important because it allows you to circumvent the restriction that a return statement can return only a single value as output. Suppose you want to calculate both the area and circumference of a circle. Does this require two separate functions? No. Consider Document 6.5.

Document 6.5 (circleStuff.htm)

```
<html>
<head>
<title>Circle Stuff</title>
<script language="javascript" type ="text/javascript">
      function circleStuff(f)  {
          var r=parseFloat(f.radius.value);
          f.area.value=Math.PI*r*r;
          f.circumference.value=2.*Math.PI*r;
      }
</script>
</head>
<body bgcolor="#99ccff">
<h1>Circle Stuff</h1>
<form>
   Enter radius, then press tab key or click on "area"
```

```
box.<br />
   radius (cm):
   <input type="text" name="radius" size="6"
      maxlength="7" value="-99",
      onblur = "circleStuff(form);" />
   area (cm<sup>2</sup>):
   <input type="text" name="area" size="6"
      maxlength="7" value="-99" />
   circumference(cm):
   <input type="text" name="circumference" size="6"
      maxlength="7" value="-99" />
</form>
</body>
</html>
```

Circle Stuff

Enter radius, then press tab key or click on "area" box.

radius (cm): `3.3`   area (cm²): `34.211943` circumference(cm): `20.734511`

Document 6.5 includes an additional form field for the circumference, calculated in the

```
f.circumference.value=2.*Math.PI*radius;
```

statement in `circleStuff()`. Both the area and the circumference are calculated within the function, but no `return` statement is used.

It is not quite true that a function accepting a form name as a parameter must know the values of all the `<input... />` tag `name` attributes. Recall from Chapter 5 that all form fields are available in an array called `elements[]` that is automatically created along with a form. The following modification of the function in Document 6.5, which uses the `elements[]` array to access the form fields, will also work:

```
function circleStuff(f) {
      var r=parseFloat(f.elements[0].value);
      f.elements[1].value=Math.PI*r*r;
      f.elements[2].value=2.*Math.PI*r;
}
```

In this case, the function must still be aware of the physical meaning of each form field as well as its position among the other fields.

It is important to understand that the significance of Document 6.5 rests on its demonstration of how to use a single function to generate more than one "output" value, in order to circumvent the requirement that a function can "return" only a single value.

6.4 Some `Global` Methods and Event Handlers

6.4.1 `Global` Methods

This book has already made extensive use of the `parseFloat()` method. Table 6.1 lists several methods of the `Global` object, including `parseFloat()`.

Table 6.1. Some `Global` methods for evaluating and converting strings

`Global` Method	Descriptions and Examples
`eval("s")`	Evaluates string `"s"` as though it were JavaScript code.
	`Eval("3+4/5")` returns a value of 3.8.
`isNaN("s")`	Returns "true" if the argument **cannot** be interpreted as a number, "false" otherwise.
	`isNaN("a17")` returns a value of true.
`parseFloat("s")`	Converts a string to a real (floating point) number.
	`parseFloat("17.7")` returns a value of 17.7.
`parseInt("s",b)`	Converts a string to an integer number using base "b" arithmetic.
	`parseInt("17.7",10)` returns a value of 17.

The last two methods are particularly important because they provide a mechanism for converting the text values of form fields into numerical values. The `parseInt()` method requires additional discussion. Consider Document 6.6.

Document 6.6 (`parseIntBug.htm`)

```
<html>
<head>
<title>parseInt() "bug"</title>
</head>
<body>
<form>
```

```
integer value: <input name="x" value="09" /><br />
Click for parseInt("string") result: <input name="x_int"
  onclick="x_int.value=parseInt(x.value); " /><br />
Click for parseInt("string",10) result: <input
name="x_int10"
  onclick="x_int10.value=parseInt(x.value,10);" /><br />
Click for parseFloat("string") result:
  <input name= "x_float"
  onclick="x_float.value=parseFloat(x.value); " />
</form>
</body>
</html>
```

integer value: 09
Click for parseInt("string") result: 0
Click for parseInt("string",10) result: 9
Click for parseFloat("string") result: 9

The parseFloat() method produces the expected value, but parseInt() with a single string argument does not. Why not? The problem lies in how ParseFloat() interprets numbers. This method can accept two arguments. The first is the text that is to be converted to an integer, and the second, described as "optional" in JavaScript documentation, is the "radix," or the number base used for the conversion of the string given as the first argument. When the second argument is absent, parseInt() tries to determine the appropriate base from the string itself. Strings of digits beginning with a zero are assumed to be base-8 (octal) numbers, not base 10! In Document 6.6, an entry of "07" will not cause a problem because 7 is an allowed digit in a base-8 system. However, 8 and 9 are not allowed digits in the base-8 system, so parseInt("09") returns 0 rather than 9! This is a perfect example of behavior that some might consider a "feature," but which others might consider a very annoying bug.[1]

The behavior of parseInt() is cause for concern because it is always tempting to ignore "optional" arguments. Consider that a two-digit format is standard for entering months, days, hours, minutes, degrees, etc., and there may be good reasons for treating whole numbers as integers rather than floating point numbers.[2] For example, it is reasonable to expect users to enter November 8, 2006, as 11/08/2006 rather than

---

[1] I admit to learning about this "feature" only when someone showed me that one of my own applications gave obviously erroneous results.

[2] At least in some programming environments, integers are stored internally in a different format than floating point numbers, which has implications for mathematical operations carried out on integers.

11/8/2006. In this case, a day entered as 08 and converted to an integer
using parseInt() would have a value of 0 (or possibly "not a number")
rather than 8—a serious error! Hence, parseInt() should *always* be
called with both arguments. Without exception for the topics addressed in
this book, the second argument should be 10, to force conversion to a
base-10 integer even with one or more leading zeros. (The first exercise
for this chapter suggests another solution, but this is intended just as an
exercise in using String methods, rather than as the preferred solution to
this parseInt() "bug.")

For the purposes of this book, where the examples are really intended
more for "local" than "global" use, it is probably not worth the effort to check
the validity of all entries in fields that are supposed to be numbers. The
isNaN() method provides a way to do this, but it has some limitations.
Referring to Document 6.6, we can see that isNaN(parseInt(x.value))
would return a value of "false" for the default entry of 09 in the "x" field
(meaning that it is a valid number) even though Document 6.6 makes it clear
that the value returned from parseInt() without the second base argument
is in error. The fact that isNaN(parseInt(x.value)) would return a
value of "true" for an entry of .09 is perhaps not so surprising, because .09 is
not an integer, but it might be a misleading result.

The eval() method listed in Table 6.1 is very powerful, and it is
worth looking online for more information about its use. Document 6.7
shows how to use the eval() method to implement a very simple
calculator that recognizes the four basic arithmetic operators (+, −, *, and /)
and knows how to interpret parentheses. The same approach can also be
used to evaluate much more complicated expressions—basically anything
that can be interpreted as one or more JavaScript statements.

Document 6.7 (calculator.htm)

```
<html>
<head>
<title>Simple Calculator</title>
</head>
<body>
<form>
   Type expression to be evaluated, using numbers
      and +, -, *, /:
   <input type="text" name="expression" size="30"
      maxlength="30"
      onchange="result.value=eval(expression.value);"
   />
   <input type="text" name="result" size="8"
```

```
     maxlength="8" />
</form>
</body>
</html>
```

Type expression to be evaluated, using numbers and +, -, *, /:

| 3*(5-13.5)/17 | -1.5 |

6.4.2 Using Event Handlers with Forms and Functions

JavaScript is an event-driven language, meaning that scripts are activated as a result of events that happen in an HTML form. The `onblur` event handler was first used in Document 6.1 and `onclick` in Chapter 4. Whenever a user tabs to a form field or clicks on that field with a mouse, that field is said to be **in focus**. The `onblur` event handler initiates JavaScript activity whenever a document user presses the Tab key to leave a particular form field or clicks elsewhere on a document. Several event handlers that can be used in this way are summarized in Table 6.2. Note the spelling of the names using only lowercase letters. These are the "official" names, and the convention should be followed even though spellings using some uppercase letters (e.g., `onBlur` is common) will be accepted by case-insensitive HTML.

Table 6.2. Summary of some event handlers used in forms

Event Handler	Action
onblur	Initiates action when a user tabs from a form field or clicks elsewhere in a document.
onchange	Initiates action when a user changes the contents of a form field.
onclick	Initiates action when a user clicks on form input field.
onfocus	Initiates action when a user tabs to or clicks on a form field.

The primary use for these event handlers is to execute code that will perform operations on field values, including changing those values.

6.5 Recursive Functions

There is an important class of calculations that can be implemented with **recursive algorithms**. A standard example is the factorial function $n!$, which is defined for nonnegative integer values of n and which is equal to $n \cdot (n-1) \cdot (n-2) \ldots (1)$. For example, $5! = 5 \cdot 4 \cdot 3 \cdot 2 \cdot 1 = 120$. This function can be defined as

$n! = 1$ for $n = 1$ or $n = 0$
$n! = n \cdot (n - 1)!$ for $n > 1$

This is a recursive definition, in which $n!$ is defined in terms of $(n - 1)!$.
        Like many other modern programming languages, JavaScript supports **recursive functions**—functions that call themselves. Document 6.8 demonstrates a recursive function that calculates $n!$.

Document 6.8 (`factorial2.htm`)

```
<html>
<title>Calculate n!</title>
<body bgcolor="#99ccff">
<script language="JavaScript" type="text/javascript">
      function nFactorial(n)  {
             if (n<=1) return 1;
             else return n*nFactorial(n-1);
      }
</script>
</head>
<h1>Calculate n factorial (n!)</h1>
<p>
<form>
  Enter n (a nonnegative integer):
  <input type="text" name="n" size="2" maxlength="3"
value="0"
  onblur="form.factorial.value=
    nFactorial(parseInt(form.n.value));" />
  (Press Tab to get n!.)<br>
  <input type="text" name="factorial" size="10"
    maxlength="11" value="1" /> <br />
</form>
</body>
</html>
```

[JavaScript Application]

⚠ 4.35890006464352 is the square root of 19.

OK

The shaded line in the code contains the critical statement, in which the function calls itself. For certain mathematical functions, such as $n!$, the structure of the recursive function is easy to see from the function's mathematical definition. Recursive algorithms always require at least two branches: one to generate a recursive call and the other to terminate the function. In Document 6.8, the relationship between the recursive definition for $n!$ and the code required to evaluate $n!$ should be obvious. Note that the code does not check to make sure that only nonnegative integer values of n have been entered as input.

The success of recursive functions depends on the function model discussed at the beginning of this chapter, in which information flows into a function through the parameter list. When the function is called with the current value of $n - 1$, this value is associated with the parameter n in the new call. Owing to the way in which the algorithm is written, the local value of $n - 1$ will eventually equal 1 (for any value of n originally greater than 1) and the recursive calls will be terminated. The intermediate values of the factorial function are stored within the programming environment. Table 6.3 shows the sequence of events for calculating 4!.

Referring again to our earlier analogy, you can think of each function call as adding a plate to a stack of plates. The initial call plus the three recursive calls add a total of four plates to the stack. As a result of the third recursive call, $n = 1$ and a value of 1 is returned. Executing a `return` statement is equivalent to removing one of the plates. Subsequently, the three remaining plates are removed as the deferred multiplications are carried out and a value is returned. When the function returns control of the script back to the point from which it was initially called, it is as though all the plates have been removed from the stack.

Table 6.3. Calculating 4! using a recursive algorithm

Local Value of n	Action	Value Returned
$n = 4$	Initial call	Deferred
$n = 3$	1st recursive call	Deferred
$n = 2$	2nd recursive call	Deferred
$n = 1$	3rd recursive call	1
$n = 2$	Complete multiplication 2·1	2
$n = 3$	Complete multiplication 3·2	6
$n = 4$	Complete multiplication 4·6	24

For more complicated recursive algorithms, it can be difficult to actually follow the course of the calculations, but fortunately, it is not necessary. As long as the algorithm is properly designed, with a condition that will eventually terminate the recursive calls, the programming environment takes care of keeping track of all the intermediate values generated during the execution of the algorithm.

Another example of a well-known function that is defined recursively is shown below. The Fibonacci numbers F_n that form the sequence 1, 1, 2, 3, 5, 8, 13, 21,... are defined for positive integer values of n as

$F_n = 1$ if $n = 1$ or $n = 2$
$F_n = F_{n-1} + F_{n-2}$ if $n > 2$

Document 6.9 shows how simple it is to evaluate this function using a recursive algorithm.

Document 6.9 (fibonacci.htm)

```html
<html>
<title>Calculate Fibonacci numbers</title>
<body bgcolor="#99ccff">
<script language="JavaScript" type="text/javascript">
   function Fib(n) {
      if (n<=2) return 1;
      else return Fib(n-1)+Fib(n-2);
   }
</script>
</head>
<h1>Calculate the n<sup>th</sup> Fibonacci number</h1>
<p>
<form>
   Enter n (a positive integer):
   <input type="text" name="n" size="2" maxlength="3"
value="1"
   onblur="FibN.value=Fib(parseInt(n.value));" />
(Press Tab to get n<sup>th</sup>
   Fibonacci number.)<br>
   <input type="text" name="FibN" size="8"
      maxlength="8" value="1" />
</form>
</body>
</html>
```

Calculate the nth Fibonacci number

Enter n (a positive integer): $\boxed{8}$ (Press Tab to get nth Fibonacci number.)
$\boxed{21}$

This function requires multiple recursive calls, and it is not easy to follow the sequence of events. However, you do not have to worry about these details as long as the algorithm is written properly!

Recursive algorithms can also be formulated using count-controlled or conditional loop structures. However, a recursive formulation is often much shorter and more direct to implement in code. The famous "Towers of Hanoi" problem is an excellent example of a problem that is difficult to solve "directly" but is trivial to solve recursively.

> Consider three poles, on one of which are stacked 64 golden rings. The bottom ring is the largest and the others decrease in size. The object is to move the 64 rings from one pole to another, using the remaining pole as a temporary storage place for rings. There are two rules for moving rings:
>
> 1. Only one ring can be moved at a time.
> 2. A ring can never be placed on top of a smaller ring.
>
> Describe how to move the entire stack of rings from one pole to another.

It can be shown that it will take $2^n - 1$ moves to move n rings. For $n = 64$, if you could move one ring per second without ever making a mistake, it would take roughly 100 times the estimated age of the universe! However, we can develop an algorithm that will work, in principle, for any number of rings and apply it to a value of n that is small enough to be practical. For $n = 4$, it will take 15 moves.

In a conceptual sense, the solution is easy (but perhaps not obvious). Suppose the poles are labeled A, B, and C. Initially, all the rings are on A and the goal is to move them all to C. The steps are:

1. Move $n - 1$ rings from A to B.
2. Move the nth ring from A to C.
3. Move $n - 1$ rings from B to C.

This solution is "conceptual" in the sense that we have not yet specified how to carry out steps 1 and 3; only step 2 defines a specific action that can be taken. However, the power of recursive functions allows us to solve this problem without giving additional specific steps! Consider Document 6.10.

Document 6.10 (towers.htm)

```
<html>
<head>
<title></title>
<script language="javascript" type="text/javascript">
  function move(n,start,end,intermediate) {
   if (n > "0") {
    move(n-1,start,intermediate,end);
    document.write("move ring "+n+
      " from "+start+" to "+end+".<br />");
    move(n-1,intermediate,end,start);
    }
   }
  var n=prompt("Give n:");
  move(n,"A","C","B");
</script>
</head>
<body>
</body>
</html>
```

move ring 1 from A to B.
move ring 2 from A to C.
move ring 1 from B to C.
move ring 3 from A to B.
move ring 1 from C to A.
move ring 2 from C to B.
move ring 1 from A to B.
move ring 4 from A to C.
move ring 1 from B to C.
move ring 2 from B to A.
move ring 1 from C to A.
move ring 3 from B to C.
move ring 1 from A to B.
move ring 2 from A to C.
move ring 1 from B to C.

Amazingly, this simple "conceptual" code is all that is required to solve this problem in the sense that all the steps are explicitly written out. Do not try this code with large values of n!

The success of this algorithm depends, once again, on how parameter lists work—passing information along a "one-way street" into a function. In principle, you can manually follow the individual values of the parameters during the recursive calls, but it is hardly worth the effort. All that is actually required is that the algorithm be stated appropriately.

6.6 Passing Values from One Document to Another

Just as it is useful to be able to pass values to functions within an HTML document, it might be useful to be able to pass values from one document to another. A typical problem is as follows:

> Create a "sign on" page that asks a user for an ID and password. Check the values provided and, if they are OK, provide access to a second page. Otherwise, access to the second page will be denied. The second page will be able to make use of information about the user that can be accessed through the user's ID.

JavaScript is not actually a suitable language for solving this problem because of the lack of two-way interaction between the client and the server. This means, essentially, that a list of approved IDs and passwords must be sent to the client computer—not a great idea! (You can "hide" this information in a separate file, as described in Chapter 5, but this is still not a real solution.) Nonetheless, it is still interesting to see how to pass information from one document to another. Document 6.11 provides a simple example.

Document 6.11a (`passID.htm`)

```html
<html>
<head>
<title>Get ID and password.</title>
<script language="javascript" type="text/javascript">
   function checkIDPW() {
     var PWinput=login_form.PW.value;
     var IDinput=login_form.ID.value;
     var flag=prompt("ID = "+IDinput+
       ", PW = "+PWinput+". OK (y or n)?");
     if (flag == "y") return true; else return false;
   }
</script>
</head>
<body>
  <form method="link" action="catchID.htm"
    name="login_form" onsubmit="checkIDPW();">
  ID: <input type="text" name="ID">
  PW: <input type="text" name="PW">
```

```html
   <input type="submit" value="Access protected page.">
</form>
</body>
</html>
```

Document 6.11b (catchID.htm)

```html
<html>
<head>
<title>Receive ID and password from another
  document.</title>
</head>
<body>1<form name="catchForm">
<input type="hidden" name="info">
</form>
<script language="javascript" type="text/javascript">
catchForm.info.value=window.location;
// alert(window.location);
function getID(str)
{
   theleft=str.indexOf("=")+1;
   theright=str.lastIndexOf("&");
   return str.substring(theleft,theright);
}
function getPW(str) {
   theleft=str.lastIndexOf("=")+1;
   return str.substring(theleft);
}
document.write("ID is "+getID(catchForm.info.value)+
   ", PW is "+getPW(catchForm.info.value));
</script>
</body>
</html>
```

Document 6.11a is the "sign on" page. It asks the user for an ID and password. The form uses method="link" to submit data to another document—catchID.htm. Since no additional location information is given, the second document must reside in the same directory (or folder) as the first one. When the link is made to the second form, the first form provides a text string that can be accessed as window.location. This consists of the URL of the first form plus values of all the form fields defined in the first document. If you know the format of this string, it is possible to extract the form field values—in this case, an ID and password.

In Document 6.11b, methods of the `String` object are used to extract substrings of `window.location`. By removing the comment characters from the `alert(window.location);` statement, you can see the entire string and how it is formatted.

This code requires that there be no "surprises" in the ID and password values. Their contents should be restricted to letters and digits. Other characters may be translated into their hex code representations, which will complicate their extraction from `window.location`. Although it might be possible, in principle, to extract several passed values, using more values will complicate the code.

Although it was not done in Document 6.11b, the implication of the code is that you can save the ID and password by assigning them to the value of a form field in the new document. Then you can use these values just as you would any value created directly within this document.

6.7 Revisiting the JavaScript `sort()` Method

Recall Document 5.3, which introduced JavaScript's `sort()` method. That example demonstrated that the results are browser-dependent and therefore unreliable. In at least some browsers, `sort()` treats array elements that "look" like numbers as though they were characters. Thus, 13 is less than 3 in the same sense that "ac" is less than "c." To fix that problem, you have to create a separate function that is passed as a parameter to the `sort()` method. This function should accept as input two values x and y (elements in the array being sorted) and should return a value of -1, 0, or 1 depending on whether x is less than, equal to, or greater than y.

In this way, you can provide your own code for comparing values. For example, you can transform text to actual numbers so that 13 will be greater than 3. Consider the following modification of Document 5.3:

Document 6.12 (`sort2.htm`)

```
<html>
<head>
<title>Sorting Arrays</title>
<script language="javascript" type="text/javascript">
  function compare(x,y) {
    var X=parseFloat(x); Y=parseFloat(y);
      if (X<Y) return -1;
      else if (X==Y) return 0;
```

```
      else return 1;
   }
   var a=[7,5,13,3];
   var i;
   document.write(a + " length of a = " + a.length+"<br />");
   a.sort(compare);
   document.write(a + " length of a = " + a.length+"<br />");
</script>
</head>
<body>
</body>
</html>
```

The two calls to document.write() in Document 6.12 show the array before and after sorting; it is clear that this code works as expected. Your "compare" function can have

| 7,5,13,3 length of a = 4 |
| 3,5,7,13 length of a = 4 |

any name you choose, as long as you use it consistently. The general idea is that, in order to force JavaScript to sort an array correctly, you have to do appropriate data type conversions in the "compare" function along with type-appropriate comparisons.

6.8 More Examples

A thorough understanding of how functions and methods work is essential to using HTML/JavaScript as a reliable problem-solving environment. As described earlier in Section 6.3, there are several different approaches to getting information to and from a function. By design, the problem to be solved in these earlier examples—calculating the area and/or circumference of a circle—was conceptually trivial. The purpose of the solutions presented was to provide templates that you can adapt for use in your own code. When JavaScript code does not work, the reason is often that a function interface has been implemented incorrectly. Hopefully, the examples presented in this section will provide some points of reference for your own code.

Example 6.1

The dewpoint temperature is the temperature at which water vapor condenses from the atmosphere. It is related to air temperature and relative humidity through the following equations:

$$a = 17.27$$
$$b = 237.7$$
$$\alpha = aT_a/(b + T_a) + \ln(RH)$$
$$T_{dp} = (b + \alpha)/(a - \alpha)$$

where relative humidity RH is expressed as a decimal fraction (between 0 and 1) and air and dewpoint temperatures T_a and T_{dp} are in degrees Celsius.

Document 6.13 (dewpoint.htm)

```
<html>
<head>
<title>Dewpoint Calculator</title>
<body bgcolor="#99ccff">
<script language="JavaScript" type="text/javascript">
  function getDewpoint(T,RH) {
    var a=17.27,b=237.7,alpha;
    var temp=parseFloat(T.value);
    var rh=parseFloat(RH.value)/100.;
    alpha=a*temp/(b+temp)+Math.log(rh);
    return Math.round(b*alpha/(a-alpha)*10.)/10.;
  }
</script>
</head>
<h1>Dewpoint Temperature Calculator</h1>
<p>
<form>
<input type="reset" value="Reset" /><br />
Temperature:
<input type="text" name="T" size="5" maxlength="6"
  value="-99" /> °C <br />
Relative Humidity:
<input type="text" name="RH" size="6" maxlength="6"
  value="-99" /> % <br />
<br />
<input type="button"
  value=    "Click here to get dewpoint temperature
(deg C)."
  onclick="DP.value=getDewpoint(T,RH)" />
<br /><br />
Dewpoint Temperature: <input type="text" name="DP"
size="5" maxlength="6" value="-99" /> °C<br />
</p>
</form>
</body>
</html>
```

Dewpoint Temperature Calculator

Reset

Temperature: 29   °C
Relative Humidity: 83   %

Click here to get dewpoint temperature (deg C).

Dewpoint Temperature: 25.8   °C

It is not absolutely necessary to define the local variables a, b, and alpha in function getDewpoint(), but it makes the conversion of the equations into JavaScript easier to understand. Note the use of the toFixed() method to control the display of the result.

Example 6.2

Given the principal P of a loan, an annual interest rate R in percent, and a repayment period of n months, the monthly payment M is given by:

$$r = R/(100 \cdot R) \quad M = (P \cdot r)/[1 - 1/(1 + r)^n]$$

create an HTML document that asks the user to enter P, R, and n and then calculates and displays the monthly payment.

Document 6.14 (loan.htm)

```
<html>
<head>
<title>Loan Calculator</title>
<body bgcolor="#99ccff">
<script language="JavaScript" type="text/javascript">
    function getPayment(P,r,n) {
        r=r/100./12.;
```

```
            var M=P*r/(1.-1./Math.pow(1.+r,n));
            return M.toFixed(2)
         }
</script>
</head>
<h1>Loan Calculator</h1>
<p>
<form>
Principal Amount: $:
<input type="text" name="amount" size="9"
  maxlength="9" value="0" /><br />
Annual rate: %
<input type="text" name="rate" size="6"
  maxlength="6" value="0" />
<br />
Number of Months:
<input type="text" name="n" size="3"
  maxlength="3" value="0" /><br />
<input type="button"
  value="Click here to get monthly payment."
  onclick=
    "monthly.value=getPayment(parseFloat(amount.value),
     parseFloat(rate.value),parseInt(n.value,10));" />
<br />
Monthly Payment: $
<input type="text" name="monthly" size="9"
  maxlength="9" />
</form>
</body>
</html>
```

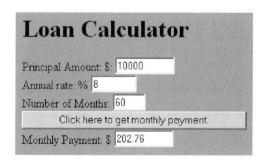

Example 6.3

As noted previously, the first application of JavaScript was to check a user's entries in form fields. Consider this problem.

> Create an application that asks a user to enter a date, using numerical values, and then calculates the day of the year. The possibilities for inappropriate entries include entering a month greater than 12 or a day number that is too large for a particular month (February 29 in a nonleap year). Prior to doing the day of the year calculation, JavaScript should check for input errors and alert the user when it detects a problem.
>
> An algorithm for finding the day of the year n based on the four-digit year, month, and day is
>
> $$n = <275m/9> - <(m + 9)/12 > (1 + <(\text{mod}(y,4) + 2)/3>) + d - 30$$
>
> where "$<...>$" means "the truncated integer value of...." For example, $<11/3> = 3$. "mod" is the remainder from integer division. For example, $\text{mod}(11/3) = 2$.
>
> This formula is valid for any year, including leap years, except for those centurial years that are not evenly divisible by 400. Thus the formula applies to 2000, which is a leap year, but not to 1900 or 2100, which are not leap years even though they are evenly divisible by 4.

Start with an outline of the code and the `form` interface.

```html
<html>
<head>
<title>Day of Year</title>
<script language="javascript" type="text/javascript">
  function checkLeapYear(y) {
  }
  function checkMonth(m) {
  }
  function checkDay(d) {
  }
  function getDayNumber(m,d,y) {
  }
</script>
</head>
<body>
<form>
  Please enter a date here in mm/dd/yyyy format:<br />
  month (1-12) <input />
  day (1-31, as appropriate) <input />
  year (20xx) <input /><br />
  The day number is <input /><br />
</form>
</body>
</html>
```

It is left as an exercise to complete this code. Here is a hint: you can make good use of the `switch` construct to check the input month and day values.

Example 6.4

In an earlier introduction to creating pull-down menus with the `select` tag (see Document 3.4), the options in the list were "hard coded" into the HTML document using the `option` tag. It is also possible to let JavaScript create the menu entries using an array of items. Document 6.15 illustrates how to do this.

Document 6.15 (`buildMenu.htm`)

```
<html>
<head>
<title>Build a variable-length pull-down menu</title>
<script language="javascript" type="text/javascript">
  var listItems = new Array();
  listItems[0]="thing1";
  listItems[1]="thing2";
  listItems[2]="thing3";
  listItems[3]="things4";
  listItems[4]="newthing";
  function buildSelect(list,things) {
    var i; //alert(things);
    for (i=0; i<things.length; i++)
      list.options[i]=new Option(things[i],things[i]);
  }
  function getSelected(list) {
    var i;
    for (i=0; i<list.length; i++)
      if (list.options[i].selected)
        return list.options[i].value;
  }
</script>
</head>
<body onload="buildSelect(menuForm.stuff,listItems);" >
<form name="menuForm" >
Here's the menu:<br />
Click on an item to select it.<br />
<select name="stuff" size="10"
  onchange="listChoice.value=getSelected(stuff);">
</select><br />
This is the item selected:
<input type="text" name="listChoice" value=" "/>
</form>
</body>
</html>
```

The list of menu items is created as an array and copied into the `options` array associated with the pull-down list. The `options` array is a property of the `select` tag (don't try to change its name) whose elements contain all the `option` tags to be defined within the `select` tag. This operation uses the `new Option()` constructor for the options array:

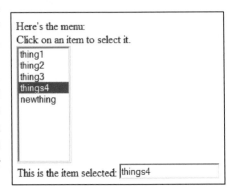

new option identifier = `new Option` (*text to appear in options list,*
text assigned to value attribute);

In Document 6.15, the process of building the pull-down menu is initiated automatically when the document is loaded on the user's computer through the `onload` event handler appearing in the `<body>` tag. The selected item is displayed whenever a choice is made in the pull-down menu by using the `onchange` event handler inside the `<select>` tag.

The text to be displayed in the pull-down menu can be the same as the text assigned to the `value` attribute for the `option` tag (as it is in Document 6.15), but it does not have to be the same. If the two input parameters for the `new Option()` constructor are different, then you need two arrays to generate these values, or perhaps a simulated two-dimensional array, as discussed in Chapter 5.

As Document 6.15 demonstrates, you do not have to "hard code" any of the `option` tags within the `select` tag. You could also just define the `options` array elements directly instead of assigning them indirectly through an additional array of items. The point of using this additional array is that you can maintain an array of menu options in another file, which can be pasted into your script as needed. In fact, this array can be a simulated two-dimensional "data array" that contains many additional values for each entry into the pull-down menu. After the user selects an item in the menu, additional form fields can be populated with information contained in the selected element of the data array.

This approach might be worth using for a long list of menu items in order to make the body of the HTML shorter and easier to read even if it is a "static" list that does not have to be changed,. Document 6.15 includes a function showing how to determine which item in the list has been selected. It uses a `for...` loop rather than a conditional loop because it is also possible in principle to specify multiple selections in a pull-down list. (Consult an HTML or JavaScript reference manual for details.)

Glossary

Glossary entries are highlighted in **bold font** at their first appearance in the text. The reference in parentheses gives the chapter and section where the word or phrase first appears.

ANSI (Preface)
American National Standards Institute, an organization that sets voluntary standards and definitions in a number of scientific and engineering areas, including computer programming languages.

array (5.1)
A collection of related elements referenced by a common name and accessed by indices.

attribute (1.2)
A value used inside an HTML element for the purpose of assigning properties and values to that element.

branching structure (4.1)
A structure that determines which section of code will be executed, based on evaluating and comparing values of one or more control variables.

calling argument (1.2)
A value passed to a method or function.

cascading style sheet (2.6)
A format for specifying attributes and values for certain HTML elements in a way that makes it easy to apply the same attributes and values throughout a document, or across many documents.

class name (2.6)
As applied to style sheets, a name by which a style definition can invoked within an element.

client-side application (1.1)
An application that resides on a user's (client's) computer without giving that user access to the host (server-side) computer.

compiled programming language (1.1)
A programming language in which one or more separate applications translate coded statements into a separate file that can then be used to execute the program.

conditional loop (4.8)
A loop structure whose operation and termination is governed by values generated inside the loop while it is executing.

constructor (5.1)
A method used to create new instances of an object.

count-controlled loop (4.8)
A loop structure whose operation and termination is governed by an index whose beginning and terminating values are specified ahead of time.

data declaration (4.4)
The process by which a variable is given a name and, optionally, assigned a value.

data type (4.4)
A definition for information stored, accessed, and manipulated in a specific way.

element (1.1, 5.1)
A construct for creating and controlling content in an HTML document. In JavaScript, the contents of one entry in an array.

escape sequence (1.2)
A way of displaying characters that are not available on the keyboard or that would be misinterpreted by HTML if entered directly. For example, `&lt;` is the escape sequence for the < symbol, which HTML interprets as the beginning of a tag if it is entered directly from the keyboard.

event handler (4.9)
An HTML attribute that initiates a response to certain user actions on a Web page, such as moving a cursor over a particular form field.

field (3.2)
A component of a form that, through an `input` element, allows user input, displays the results of calculations, or provides controls over form processing.

floating point (3.1)
A method of storing real numbers, associated with "floating point" (as opposed to integer) variables.

form (3.1)
An HTML element that provides an interface between a user and a document. The contents of form fields can (usually) be changed by the user or by a scripting language within the document.

free-format language (4.3)
A language such as JavaScript that, within syntax limits, does not restrict where and/or how statements are placed on a document line.

function (6.1)
A self-contained code unit that accepts input, performs one or more specific tasks, and returns output.

hex code (2.5)
A number expressed in hexadecimal notation, using values 0 through F (rather than 0 through 9, as in a base-10 number system).

home page (1.3)
The top level Web page associated with a Web address. By default, the HTML document containing the home page is called `index.htm` or `index.html`.

HTML (document) (Preface)
H̲yper̲T̲ext M̲arkup L̲anguage, a language for displaying and accessing online content, especially on the World Wide Web. HTML is approximately, but not entirely, platform-independent, as different browsers support different subsets and extensions of HTML. An HTML document is any `.htm` or `.html` text file that uses HTML for organizing and displaying text, images, and other content.

HTTP (1.3)
H̲yper̲T̲ext T̲ransfer P̲rotocol, a communications protocol for exchanging information on the World Wide Web.

in focus (6.4)
The state of a form field or other defined area on a Web page when the cursor is within that field or area.

identifier (4.4)
A symbolic name associated with a variable.

indices, (index) (5.1)
One or more values that identify a single array element.

input/outut (I/O) interface (1.1)
A system that manages interactions between a user and a document, program, or script.

Internet (1.3)
A globally connected network of computers for exchanging information using an agreed-upon communications format.

intranet (1.3)
A system of linked computers that looks like the Internet, but is accessible only to other computers on an internal network.

interpreted programming language (1.1)
A language, such as JavaScript, in which statements are interpreted one line at a time, and the indicated actions taken "on the fly," without generating a separate executable file. (See **compiled programming language**.)

JavaScript (Preface)
An object-oriented programming language designed for manipulating content in an HTML document.

list (3.6)
In HTML, one of several ways to impose formatting on lists (in the plain English use of that word) of related items.

literal (4.4)
A value entered directly in code, rather than being associated with a variable.

local variable (6.2)
A variable defined inside a function that is visible only within that function.

logical operator (4.7)
An operator that determines whether two expressions are both true ("AND"), one of two expressions is true ("OR"), or neither of two expressions is true ("NOT").

loop structure (4.1)
A code structure that enables a section of code to be executed more than once, under the control of one or more index or control variables.

method (1.2)
An action that can be applied to an object or components of an object.

object (1.3)
A defined construct that has components, properties, and values, and that allows certain actions to be carried out upon itself or its components.

object-oriented programming language (1.1)
Any language that makes used of objects.

operator (4.5)
A token representing a mathematical, logical, or text action, such as addition.

parameter list (6.2)
A list of references to one or more values passed as input to a function.

platform-independent (Preface)
A computer language or application that presents a uniform user interface and behavior regardless of the computer or operating system.

post-test loop (4.8)
A conditional loop structure in which tests for termination or continuation are conducted at the end of the loop's statement block.

precedence rules (4.5)
The rules governing the order in which operations, including mathematical and logical operations, are performed in a statement.

pre-test loop (4.8)
A conditional loop structure in which tests for termination or continuation are conducted at the beginning of the loop's statement block.

primitive (4.4)
A basic data type.

property (1.2)
An attribute (in the plain English use of that word) of an object or one of its components.

queue (5.2)
An abstract data type, often represented by an array, in which the first entry (the "oldest" entry at the "front" of the queue) is always the first to be removed and new entries are always added at the opposite end ("back") of the queue.

recursive algorithm (6.5)
An algorithm that depends on being able to refer to itself.

recursive function (6.5)
A function that refers to itself by calling itself.

relational operator (4.7)
An operator that compares the value of two expressions and returns a value of true or false.

script (1.2)
The statements ("code") used to implement a scripting language such as JavaScript.

scripting language (1.1)
A language such as JavaScript whose purpose is to access and modify components of an existing information interface.

stack (5.2)
An abstract data type, often represented by an array, in which the entries are always added to the "top" of the stack and the most recent entry is always the first to be removed.

statement 4.3)
A single set of instructions, often followed by a terminating character (a semicolon in JavaScript).

statement block (4.3)
Several statements meant to be treated as a group, marked with a special symbol at the beginning and end of the block, for example, { ... } in JavaScript.

style rule(s) (2.6)
One or more attributes and values defined within a style sheet.

style sheet (2.6)
(See **cascading style sheet**.)

table (3.1)
An HTML element that provides a way to organize and display content in a document.

tag (1.1)
A syntax for entering elements in HTML documents (<...>), usually involving both a start tag and an end tag.

terminating character (4.3)
A character appearing at the end of a programming statement, to mark the end of the statement. The JavaScript terminating character is a semicolon.

token (4.2)
The smallest, indivisible, lexical unit of a programming language. Tokens can be constants, identifiers, operators, reserved words, or separators (or terminators).

URL (1.3)
Uniform Resource Locator, the address system used by the World Wide Web.

variable (4.4)
A discrete unit of information, associated with a particular data type and stored in a specific part of computer memory.

weakly typed language (4.4)
A language that permits variables to be retyped (redefined) "on the fly" based on their contents and/or allows variables to be used without an initially specified data type.

Web browser (1.1)
A computer application designed to access, display, and interpret online content.

Web server (1.3)
A computer connected to the Internet that stores documents and other content for global (but not necessarily public) access by way of a unique address (Uniform Resource Locator, or URL).

World Wide Web, WWW (1.1)
A global network of computing resources that uses the hypertext transfer protocol (HTTP) to exchange information on the Internet.

XHTML (1.2)
EXtended HyperText Markup Language, a more rigorous version of HTML that vigorously enforces syntax and style rules.

Appendices

A.1 HTML Document Examples

Document and Name   Page

A.2 Displaying Special Characters in an HTML Document

There are many symbols that cannot be entered directly into an HTML document. HTML defines so-called "escape sequences" as a way to embed special characters in a document. Each character can be entered either as a numerical code or by using a mnemonic name, butonly the names are used here. The following list provides some commonly used characters that may be useful for science and engineering applications. The list is a *very* small subset of characters supported by various browsers. In cases where special character names follow a predictable pattern (for the Greek alphabet, for example), just one example is given. (See notes at the end of the list.) There is no guarantee that the escape sequence names will be recognized or that characters will be displayed properly in all browsers or, when printed, by all printers.

&alpha;	α	lowercase Greek alpha[*]
&asymp;	≈	mathematical "almost equal to" symbol
&aacute;	á	lowercase "a" with acute accent[**]
&acirc;	â	lowercase "a" with circumflex[**]
&aelig;	æ	lowercase "ae" ligature (Æ for uppercase)
&agrave;	à	lowercase "a" with grave accent[**]
&aring;	å	lowercase "a" with ring[**]
&auml;	ä	lowercase "a" with umlaut[**]
&bull;	·	small "bullet" symbol (to indicate multiplication, for example)
&ccedil;	ç	lowercase "c" with cedilla[**]
&cent;	¢	cent symbol
&cong;	≅	mathematical "approximately equal to" symbol
&copy;	©	copyright symbol
&deg;	°	degree (as with temperature)
&dagger;	†	dagger symbol
&Dagger;	‡	double dagger symbol
&divide;	÷	mathematical "divide by" symbol
&euro;	€	Euro currency
&frac12;	½	fraction notation for 1/2
&frac14;	¼	fraction notation for 1/4
&frac34;	¾	fraction notation for 3/4
&ge;	≥	mathematical "greater than or equal to" symbol
&gt;	>	mathematical "greater than" symbol (to avoid conflict with angle bracket used in HTML tags)

`&hellip;`	…	horizontal ellipsis
`&infin;`	∞	mathematical "infinity" symbol
`&int;`	∫	mathematical "integral" symbol
`&iques;`	¿	inverted question mark
`&ldquo;`	"	left double quote ("smart quote")
`&lsquo;`	'	left single quote ("smart quote")
`&le;`	≤	mathematical "less than or equal to" symbol
`&lt;`	<	mathematical "less than" symbol (to avoid conflict with angle bracket used in HTML tags)
`&micro;`	μ	micron
`&ne;`	≠	mathematical "not equal to" symbol
`&ntilde;`	ñ	lowercase n with tilde[**]
`&oelig;`	œ	lowercase "oe" ligature (`&OElig;` for uppercase)
`&para;`	¶	paragraph symbol
`&plusmn;`	±	mathematical "plus-minus" symbol
`&pound;`	£	British pound sterling
`&prop;`	∝	mathematical "proportional to" symbol
`&quot;`	"	quote symbol (e.g., for inserting quote marks in quote-delimited text string)
`&radic;`	√	mathematical "square root" symbol
`&rdquo;`	"	right double quote ("smart quote")
`&rsquo;`	'	right single quote ("smart quote")
`&reg;`	®	product registration symbol
`&sect;`	§	section symbol
`&szlig;`	ß	"sz" ligature (lowercase only)
`&times;`	×	mathematical "times " ("multiply by") symbol
`&trade;`	™	trademark symbol

[*] Other Greek letters can be displayed by spelling the name of the letter. If the name starts with an uppercase character (for example, `&Gamma;`) then the uppercase letter is displayed. Otherwise the lowercase character (for example, `&gamma;`) is displayed.

[**] Other modified letters follow the same pattern. Start the name with an uppercase or lowercase letter to display a modified uppercase or lowercase character.

Exercises

1. Introductory Topics

1.1. Create a folder on whichever computer you will use with this book. Then create a simple Web page for yourself and store it in that folder. Save your home page file as `index.htm` (or `index.html`). Include the `lastModified` property to show the most recent date on which the page was modified, as in Document 1.3.

1.2. If appropriate, copy your Web page to a location where it will be available through the Internet or an intranet. If you are using this book as a course text, your instructor should provide the information you need to make your work Web-accessible.

2. HTML Document Basics

2.1. Add some content to your Web page. This can be a short biographical sketch or something less personal. Use some of the HTML elements described in this chapter. Experiment with setting different colors and font sizes. Include at least one image—preferably one you create yourself. Be sure to display the source of the image. Do not use commercial images unless you can demonstrate that you have permission to use them.

2.2. Create a style sheet file for your Web page, save it as a separate file, and modify your Web page so that it uses this style sheet. Create at least one other Web page that shares this style. The content of this second page does not matter, but there must be enough content to demonstrate that the style is being implemented. (You may want to combine this with Exercise 1.)

2.3. An internal link, essentially a "bookmark" to a specified point in a document, is created as follows:

```
<a href="#section1">Link to Section 1.</a>
```
...

```
<a name="#section1">Start of Section 1.</a>
... {text of Section 1.}
```

The # sign appearing in the value of the href and name attributes indicates that this is an internal document link. The `<a name="#...">` ... `</a>` tags typically surround a section heading, or perhaps the first few words in a section (see Section 2.4).

Create a document with a "table of contents" that is linked to several sections. At the start of each section, include a link back to the table of contents. The sections do not have to be long, as the purpose of this exercise is just to learn how to create internal document links.

2.4. Create an HTML document that contains at least two clickable images that are linked to other HTML documents. In Microsoft Word, for example, you can use the "WordArt" feature to  create graphics images that explain the link, as with these examples (see Section 2.4).

2.5. Create an HTML document that displays this heading and HTML code:

Here is some HTML code...

```
<html>
<head>
<title>Displaying HTML code in a document</title>
</head>
<body>
Here is an HTML document.
</body>
</html>
```

All the HTML tags, including their left and right angle brackets, should be displayed in red font. Note that this is not an HTML code listing. It is the displayed content of an HTML document. *Hint*: Review Document 2.1 and its explanation.

3. HTML Tables, Forms, and Lists

3.1. Create a table containing a personnel evaluation form. The first column should contain a statement, such as "Gets along well with others." The second column should have four radio buttons, showing the choices "Never," "Sometimes," "Often," and "Always." The table should have at least four statements. Provide appropriate instructions for filling in the form and submitting it to the creator of the form.

What happens if you submit the contents of a form that does not include the `enctype` attribute in its `<form>` tag? What happens if you use `method="get"` instead of `method="post"`? Show examples.

3.2. Using Table 2.1 as a guide, create an HTML document and table that displays the 16 standard HTML colors and their hex codes. The color names should be displayed in their color against an appropriate background color.

3.3. Using Table 2.2 as a guide, create a table that displays results of assigning specific and generic font families to text. For example, display an example in serif and Times fonts.

3.4. Modify Document 3.9 so that it e-mails the contents of the form to your address. What happens with the choices you have checked? What about the choices you have not checked?

3.5. Create a table containing a list of professors. The first column should contain their names and the second should allow you to send an email. If you click in the first column, a new window should open that displays information about the professor in that column.

Opening a new window has not been covered in the text. Some code to get you started is as follows:

Creating the table:

```html
<html>
<head>
<title>List of Professors</title>
</head>
<body>
<table border>
<tr><th>Biographical sketch<br />
(click in name box)</th><th>Contact</th></tr>
<tr>
```

```
<td onclick ="window.open('ProfWonderful.htm',
  'ProfWonderful','alwaysRaised=yes,toolbar=no,
  width=600,scrollbars=yes');">
  Professor Wonderful, Super University</td>
<td><a href="mailto:I.M.Wonderful@superu.edu">
  I.M.Wonderful@superu.edu</a></td>
</tr>
</table>
</body>
</html>
```

The HTML document for Professor Wonderful:

```
<head>
<title>Professor Wonderful</title>
<link href="WindowStyle.css" rel="stylesheet"
type="text/css" />
</head>
<body>
<b><i>Professor I. M. Wonderful, PhD</i></b><br />
 Enter biographical stuff about Professor Wonderful.
</body>
</html>
```

Create a `BiographyStyle.css` file that should be applied to every biography file.

This example will get you started with the `window.open()` method. You can find much more information online.

4. Fundamentals of the JavaScript Language

4.1. Prompt the user to enter a temperature in degrees Fahrenheit. Calculate and display this temperature converted to degrees Celsius and Kelvins. The conversion from Fahrenheit to Celsius is $T_C = 5(T_F - 32)/9$. The conversion from T_C to Kelvins is $K = T_C + 273$.

4.2. Prompt the user to enter two numbers. Print a message that tells which number is smaller or if they are equal. (Do not use the `Math.max()` or `Math.min()` functions.

4.3. Prompt the user to enter the temperature in degrees Fahrenheit and the wind speed V in miles per hour. Calculate and display the windchill temperature according to:

$T_{WC} = (0.279V^{1/2} + 0.550 - 0.0203V)(T - 91.4) + 91.4$

where T must be less than 91.4°F and $V \geq 4$ mph. Include code to test the input values for T and V and print an appropriate message if they are out of range.

4.4. Prompt the user to enter the month m, date d, and year. Calculate and display the day of the year n, from 1 to 365 or 366, depending on whether the year is a leap year. The formula is

$n = INT(275m/9) - k \cdot INT((m + 9)/12) + d - 30$

where INT() means "the truncated integer value of."

A year is a leap year if it is evenly divisible by 4 and, if it is a centurial year, it is evenly divisible by 400. That is, 2000 was a leap year, but 1900 was not. Provide results for several inputs, including the first and last days of leap and nonleap years, and February 28 or 29 and March 1.

4.5. Rewrite Document 4.7 so that it uses either a pre- or post-test loop. Why did you choose one conditional loop strategy over the other?

4.6. Rewrite Document 4.8 so that it uses a post-test loop.

4.7. Modify Document 4.11 so that if a user enters atmospheric pressure entered in inches of mercury instead of millibars, the code will convert that value into millibars. It is easy to distinguish such an entry because of the large difference in magnitude between the two units. Standard sea level atmospheric pressure is 1013.25 mbars or 29.921 in. of mercury. Therefore, if the user enters a numerical value less than 40, for example, it is safe to assume that it represents inches of mercury. Then, $p_{mbar}/1013.25 = p_{in.\ of\ mercury}/29.921$.

4.8. Using Document 3.1 as a starting point, let a user enter radon values into a table and form. The code should then display an appropriate message in the third column of the table, depending on the radon level.

4.9. Create a table with a form into which a user enters total credit hours and grade points for eight semesters. The code should calculate the GPA for each semester:

GPA = (grade points)/(credit hours)

where an A gives 4 credit points, a B gives 3 credit points, etc. The last line in the form should be the cumulative GPA:

cumulative GPA = (cumulative grade points)/(cumulative credit hours)

4.10. Create a table containing a price list and order form. The first column contains a brief description. The second contains the price for one item. The third column contains a form field in which the user enters the number of items to order. The fourth column contains a form field in which the extended price (price per item times number of items) is calculated. There should be data entry rows for at least three items. The table should then calculate the total amount for all items ordered, sales tax, shipping, and order total. Provide appropriate instructions for filling in the form and submitting it to the creator of the form.

4.11. The body mass index (BMI), which provides a way to characterize normal weights for human adult bodies as a function of height, is defined as:

$$BMI = w/h^2$$

where w is mass in kilograms (2.2 kg mass per pound weight) and h is height in meters (1 in. = 0.0254 m).
    Create a document and form that asks for the user's weight in *pounds* and height in *feet and inches*, and then calculates and displays the BMI.

4.12. A cylindrical liquid storage tank of radius R and length L is buried underground on its side, that is, with its straight sides parallel to the ground. In order to determine how much liquid remains in the tank, a dip stick over the centerline of the tank is used to measure the height of the liquid in the tank. The volume is $L \cdot A$, where A is the area of a partial circle of radius R with a cap cut off horizontally at height H from the bottom of the circle:

$$A = R^2 \cos^{-1}[(R - H)/R] - (R - H)(2RH - H^2)^{1/2}$$

where $\cos^{-1}(x)$ is the inverse cosine (arccosine) of x.

Create a document that accepts input values for R, L, and H and then calculates and displays the volume of liquid in the tank.

4.13. Paleontologists have discovered several sets of dinosaur footprints preserved in ancient river beds from which it is possible to deduce the speed at which these dinosaurs walked or ran. The two pieces of information that can be determined directly from the footprints are the length of the dinosaur's foot and the length of its stride, which is defined as the distance between the beginning of a footprint made by one foot and the beginning of the next footprint made by that same foot.

One way to approach this problem is to examine the relationship between size, stride, and speed in modern animals. Owing to the dynamic similarities in animal motion, an approximate linear relationship between relative stride and dimensionless speed applies to modern bipedal and quadrupedal animals as diverse and differently shaped as humans, ostriches, camels, and dogs:[1]

$$s = 0.8 + 1.33v$$

Relative stride s is defined as the ratio of stride length to leg length, $s = S/L$. Dimensionless speed is defined as the speed divided by the square root of leg length times the gravitational acceleration g, $v = V/(Lg)^{1/2}$. Although it might seem that gravitational acceleration should not influence an animal's speed on level ground, this is not true, as gravity influences the up and down motions of the body required even for walking.

Leg length from ground to hip joint for dinosaurs of a known species can be determined from fossils. However, even when the dinosaur species responsible for a set of tracks is unknown, its leg length can be estimated by multiplying the footprint length by 4. (Try this for humans.)

Create a document that uses the equation described here to calculate the speed of a dinosaur based on measurements of its footprint and stride length. Use metric units. Test your calculations for a footprint 0.6 m long and a stride length of 3.3 m.

Extra Credit Note: Is it possible to determine whether the dinosaur was walking or running? Using data for human strides—walking or running— you should be able to speculate about the answer to this question.

---

[1] See R. McNeill Alexander, *Dynamics of Dinosaurs & Other Extinct Giants.* Columbia University Press, New York, 1989.

4.14. The wavelengths of the Balmer series of lines in the hydrogen spectrum are given by

$$\lambda = 364.6n^2/(n^2 - 4) \text{ nm}$$

Write a script that generates and displays the first 10 wavelengths in the Balmer series. Use document.write() to display the results.

4.15. The original population of a certain animal is 1,000,000. Assume that at the beginning of each year (including the first year), the population is increased by 3%. By the end of that year, 6% of the total population (including the births at the beginning of the year) dies. Write a script that calculates and displays the population at the end of each year until the population at the end of the year falls to 75% or less of its original value. Although, in principle, you can figure out how many years this will take, don't do that. Use a conditional loop.

4.16. Section 4.6 briefly discussed some problems with using the Math.random() function to create a series of randomly distributed integers. Write a script that will examine the distribution of 10,000 integers in the range [0,2], using these two expressions:

```
Math.round(Math.random*2)
```

and

```
Math.floor(3*(Math.random()%1))
```

Show results from several trials. Explain your results.

5. Using Arrays in HTML/JavaScript

Each exercise in this chapter should use arrays as appropriate.

5.1. Rewrite the code from Exercise 16 in Chapter 4 so that it looks at the distribution of integers in the range [1,6]. Which of the expressions given in that exercise should be used to simulate the performance of a fair six-sided die? What would be the result of using the wrong expression?

5.2. Rewrite Document 5.9 so that it uses a conditional loop that stops searching when it finds a password match.

5.3. Rewrite Exercise 4.9 to use the `elements` array that is automatically created for the fields in a form.

5.4. The text makes the point that assigning one array name to another:

```
var A = ["thing1","thing2","thing3"];
var B = A;
```

does not actually create a separate copy of the array A. Instead, both A and B "point" to the same data in memory.

Write code that accepts the names of source and destination arrays and copies the contents of the source array into the destination array, thereby creating an actual copy of the original array. The result should allow you to manipulate the two arrays independently, without changes to one array affecting the contents of the other. Your code should display the contents of both arrays both before and after the destination array has been created and changes have been made to the contents of the source array.

5.5. Write a script that finds the maximum, minimum, mean, standard deviation, and median of an array.

mean $= \Sigma x_i / n$

(standard deviation)$^2 = [\Sigma x_i^2 - (\Sigma x_i)^2/n]/(n-1)$

where the x_i's are the elements of the array, n is the number of elements, and "Σ" means "sum from 1 through n." (Remember that in a JavaScript array, the elements are indexed by value from 0 through $n-1$, not 1 through n.)

The array must be sorted to find the median. For an array with an odd number of elements, the median is the middle value. For an array with an even number of elements, the median is the average of the two middle elements.

5.6. Write a function that will reverse the elements in an array, having the effect of turning a "stack" upside down, for example. Do not do this by creating a "backward" copy of the original array; rather make the changes to the original array. *Hint*: Start from one end of the array and work toward the middle. Create a temporary variable to hold an element at one end of the array, replace it with its corresponding value from the other end of the array, and then replace the value at the other end of the array with the temporary value.

5.7. Write a function that will create a "count histogram" for an array. To do this, specify the minimum and maximum values and the number of "boxes" into which the array should be divided. For example, you could divide an array of values into quartiles and count the number in the lowest 25%, next 25%, etc. The count histogram should be stored in its own array and the results displayed. Use this array as an example, but be sure to write your code so it will apply to any array.

```
var A=[53,97,66,79,80,81,83,75,91,65,61];
```

Your script should determine the minimum and maximum values in the array and then prompt you to provide the lower limits on the histogram and the number of subdivisions. Suppose this array contains test grades. A reasonable choice would be to divide these values into five ranges: 50–59, 60–69, 70–79, 80–90, and 90–100. The contents of the count histogram array should be 1, 3, 2, 3, 2, and the count histogram could be interpreted as showing the distribution of F's, D's, C's, B's, and A's.

Demonstrate the operation of your script for several choices of ranges and numbers of histogram "boxes."

5.8 Automata can be thought of as artificial life forms that, with the aid of a set of rules for reproducing themselves, appear to be self-organizing. These rules can lead to surprising patterns, related to fractal theory. One well-known pattern is the Sierpinski triangle, shown here.

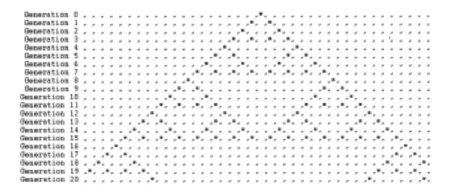

This output is generated by repeatedly printing an array with `document.write()`. In order for the elements to line up, you can specify a monospaced font style inside a `document.write()` before starting to print the output:

```
document.write("<font style='font-family:Courier'>");
```

The pattern starts out with a single "life form" (an asterisk) in the middle of the array. This array has 40 elements. The propagation rules are:

For cell i, if cell $i-1$ is occupied and cells i and $i+1$ are not, or if cell $i-1$ is empty and cell $i+1$ is occupied, then an organism will appear in cell i in the next generation. Otherwise the cell will be empty.

Write JavaScript code that reproduces the output shown. *Hint*: you cannot apply the rules to the array itself to determine the distribution of organisms in the next generation. You have to copy the organism distribution array at the start of each generation and test the propagation rules as applied to that copy in order to update the organism distribution array.

6. JavaScript Functions

These exercises should always include appropriately designed functions. Note that some of the exercises for Chapter 4 can be rewritten using functions.

6.1. Recall Document 6.5, which demonstrated how a "bug" in the `parseInt()` method can cause problems when interpreting an integer represented as a string that begins with a 0. Write a function that accepts as input an "integer" value that begins with a 0. The function should create a substring of the original value that does not have this leading 0 and should then apply `parseInt()` to this substring. (See the list of string-related methods given in Table 4.2.)

6.2. Create a document that asks a user to enter a month, day, and year. Then, check the day to make sure it is an appropriate day for that month. Don't forget about leap years, as previously defined in Exercise 4.4. Use an "alert" box to display a message if the user enters an inappropriate day.

6.3 A basic problem in numerical analysis is the solution of systems of linear equations. Consider a system of equations with three unknowns:

$$a_1x + b_1y + c_1z = d_1$$
$$a_2x + b_2y + c_2z = d_2$$
$$a_3x + b_3y + c_3z = d_3$$

Cramer's rule can be used to solve equations in two or three unknowns, but becomes unwieldy for larger systems. For the above system:

$$x = D_1/D \qquad y = D_2/D \qquad z = D_3/D$$

where D is the determinant for the system:

$$D = a_1b_2c_3 + b_1c_2a_3 + c_1b_3a_2 - a_3b_2c_1 - b_3c_2a_1 - c_3b_1a_2$$

D_1, D_2, and D_3 are found by substituting the constants d_1, d_2, and d_3 for the coefficients in column 1, 2, and 3, respectively:

$$D_1 = d_1b_2c_3 + b_1c_2d_3 + c_1b_3d_2 - d_3b_2c_1 - b_3c_2d_1 - c_3b_1d_2$$
$$D_2 = a_1d_2c_3 + d_1c_2a_3 + c_1d_3a_2 - a_3d_2c_1 - d_3c_2a_1 - c_3d_1a_2$$
$$D_3 = a_1b_2d_3 + b_1d_2a_3 + d_1b_3a_2 - a_3b_2d_1 - b_3d_2a_1 - d_3b_1a_2$$

It is possible for the value of D to be 0. Then the system of equations has no solution. Your code should test for this possibility and provide an appropriate message. Include your solution for the following system of equations:

$$3x + 4y + 2z = -1$$
$$5x + 7y + z = 2$$
$$5x + 9y + 3z = 3$$

6.4. Complete the code outlined in Example 6.3 at the end of Chapter 6. The algorithm for finding the day of the year n based on the four-digit year, month, and day is

$$n = <275m/9> - <(m + 9)/12>(1 + <(\mod(y,4) + 2)/3>) + d - 30$$

where "<...>" means "the truncated integer value of..." For example, $<11/3> = 3$. "mod" is the remainder from integer division. For example, $\mod(11/3) = 2$.

This formula is valid for any year, including leap years, except for those centurial years that are not evenly divisible by 400. Thus the formula applies to 2000, which is a leap year, but not to 1900 or 2100, which are not leap years even though they are evenly divisible by 4.

6.5. Write a modified version of Document 6.9 that demonstrates the relationship between Fibonacci numbers and the golden ratio. (It is easy to find a lot of information about this unexpected (?) relationship online.)

6.6. Document 6.12 shows how to use `parseFloat()` to force JavaScript to treat array elements as numerical values. The function returns a value of -1, 0, or $+1$. depending on whether x is less than, equal to, or greater than y. Actually, this will work if the function returns any negative value for $x < y$, 0 for $x = y$, and any positive value for $x > y$. Write a *one-line* "compare" function that returns such a result. *Hint*: the subtraction operation $x - y$ (as opposed to the addition operation $x + y$) has no interpretation if x and y are strings, and forces an implicit type conversion to numerical values.

6.7. The `Math.floor()` method, which returns the next-lowest integer below the real number given as its calling argument, truncates nonnegative numbers in the way you would expect. For example, `Math.floor(17.9)` equals 17. However, it does not truncate negative numbers. For example. `Math.floor(-17.9)` equals -18, not -17. It might be useful to have a function that simply strips away digits to the right of the decimal point— that is, a function that actually truncates a number—regardless of its sign.

Fortunately, it is easy to create your own library of methods and properties that act as extensions to the `Math` object. It is set up in the following way:

```
function Math.myMethod(x) {
   {Put code here.}

}
```

Simply give your new method a name and write the appropriate code, with a `return` statement for the desired value. You can also create new properties just by defining them. For example, if you write the statement

```
Math.myPI=5.;
```

you now can use this new property (not that it would be a good idea!) just as you would `Math.PI`.

Write a script that creates and tests a truncation method—call it `Math.trunc()`—that works regardless of whether the calling argument is positive or negative. Also, create at least one other new `Math` method, including a method, `Math.sind()`, that returns the sine of its argument expressed in degrees rather than radians. That is, `Math.sind(30.)` should give a value of 0.5.

Of course, these methods and properties exist only within the script in which they are defined, rather than in a pre-defined library of `Math` methods and properties available to any browser that supports JavaScript. However, they are treated as "real" extensions of the `Math` object in the sense that you can use them just like other `Math` methods or properties, including referring to just the method or property names inside a `with (Math)` { ... } statement block. Once you create a library of `Math` extensions, you can simply paste them into any script—either literally or by saving them in a `.js` file and referencing that file in a script.

6.8. As noted briefly in Section 6.4, the `eval()` Global method is very powerful. Shown below is an HTML document template for an application that will numerically integrate a specified function. The default function is the normal probability density function, which does not have an analytic integral. You can replace this function with any function expressed in proper JavaScript syntax. The function uses standard trapezoidal rule numerical integration. An outline of the algorithm is as follows:

Specify a function $f(x)$, lower and upper integration boundaries (a and b), and the number of equal intervals (n) into which the range ($b - a$) will be divided. The code outline for the numerical integration is:

Set sum = 0, $dx = (b - a)/n$.

for $i = 1$ to n,
$$x_1 = a + (i-1) \cdot dx$$
$$x_2 = a + i \cdot dx$$
$$\text{sum} = \text{sum} + f(x_1) + f(x_2)$$

return sum$\cdot(dx/2)$

Create an HTML document that uses a JavaScript function and the `eval()` method to implement this algorithm. Consult a probability and statistics text or online source to verify the values produced for the normal probability distribution function. In addition, test the application by entering a function that has an analytic integral.

Numerical Integrator

NOTES:
(1) Enter function using JavaScript syntax.
(2) Math methods and properties don't require "Math." in front of their names.
(3) The default function given here is the normal probability density function.
(4) This application uses Trapezoidal Rule numerical integration.

function: `exp(-x*x/2)/sqrt(2*PI)`

lower boundary: `0`   upper boundary: `0.5`

number of intervals: `100`

    Click here to integrate...

6.9. Assume that the probability of a randomly selected individual in a target population having a disease is PD. Suppose there is a test for this disease, but the test is not perfect. There are two possible outcomes from the test:

1. Test is positive (disease is present).
2. Test is negative (no disease is present).

As the test is imperfect, if the individual has the disease, result 1 is returned for only PWD (test positive, with disease) percent of the tests. That is, only PWD percent of all individuals who actually have the disease will test positive for the disease. If the individual does not have the disease, result 2 is returned only NND (test negative, no disease) percent of the time. That is, only NND percent of all individuals who do not have the disease will test negative for the disease.

Bayesian inference can be used to answer two important questions:

1. Given a positive test result, what is the chance that I have the disease?
2. Given a negative test result, what is the chance that I have the disease anyhow?

Define the following variables (assuming that PWD and NND are expressed as values between 0 and 1 rather than as percentages):

PND = positive test result, but with no disease = $(1 - \text{NND})$
NWD = negative test result, but with disease = $(1 - \text{PWD})$

P_has_disease = person has disease, given a positive test result
= (# of true positives)/(# true positives + # false positives)
= $(\text{PWD} \cdot \text{PD})/[\text{PWD} \cdot \text{PD} + \text{PND} \cdot (1 - \text{PD})]$

Probability that a person does not have the disease, given a positive test result = $1 - \text{P\_has\_disease}$

N_has_disease = person has disease, given a negative test result
= (# false negatives)/(# false negatives + # true negatives)
= $(\text{PND} \cdot \text{PD})/[\text{PND} \cdot \text{PD} + \text{NND} \cdot (1 - \text{PD})]$

The probability that a person has the disease even though the test result is negative is called a Type II error. The probability that a person does not have the disease even though the test result is positive is called a Type I error. From a treatment point of view, Type II errors are perhaps more serious because treatment will not be offered. However, it is also possible that treating for a disease that does not actually exist, as a result of a Type I error, may also have serious consequences.

As an example, consider a rare disease for which PD = 0.001, PWD = 0.99 and NND = 0.95. Then the probability that a person has the disease, given a positive test result, is:

P_has_disease = $(0.99 \cdot 0.001)/(0.99 \cdot 0.001 + 0.05 \cdot 0.999) = 0.019$

and for a negative test result:

N_has_disease = $0.01 \cdot 0.001)/(0.01 \cdot 0.001 + 0.95 \cdot 0.999) = 0.0000105$

The somewhat surprising result that the probability of having this disease is very small despite a positive result from a test that *appears* to be highly accurate is explained qualitatively by the fact that there are many more people without the disease (999 out of 1000) than there are with the disease. In such a population, approximately 50 people will test positive for the disease even though they do not have it. Approximately 1 person will test positive for the disease when they have it, so (actual positives)/(all positive test results) ~ 1/51 ~ 0.02.

The very small probability of having the disease even with a negative test result is explained by the fact that 999 out of 1000 people do not have the disease and almost all of these people get negative test results.

Write a document that displays results from the indicated calculations. What happens when the tested disease is found in 50% of the population? What happens for both disease situations when the positive and/or negative tests are much less reliable, say 50%?

6.10. Using Document 6.15 as a starting point, write an application that contains contact information for your friends and colleagues. When you select a person's name from the pull-down menu, the remaining fields in a form should be populated automatically with the contact information for that person. Store the contact information in a separate "hidden" file, to be pasted into your document when it is loaded, as shown in Chapter 5, Section 5.5. The following is a sample of what this contact information file might look like:

```
var contactList = new Array();
function contactArray(name,phone,email) {
  this.name=name;
  this.phone=phone;
  this.email=email;
}
contactList[0]=new contactArray("Mom",
  "222-555-5478","mom@supermail.net");
contactList[1]=new contactArray("My Boss",
  "888-555-0985","MrBig@xyyz.com");
contactList[2]=new contactArray("Sally",
  "111-555-2311","SallyJo@ail.com");
```

Considering previous discussions about sorting arrays in JavaScript, how should you handle the matter of keeping this list sorted in alphabetical order by name? Do you have to sort the contact list array "offline" or is it reasonable to do it in "real time" every time the pull-down menu is created?

6.11. Newton's algorithm for finding the square root of a number, as discussed in Section 4.8.2, can also be implemented (actually, more easily) as a recursive function. Rewrite Document 4.10 so that the iterative calculation is done in a function, add another function that does the same calculation recursively, and then add two functions that use a similar approach to finding the cube root of a positive real number:

1. Select an initial guess $g = n/2$.
2. Replace g with $(2g + x/g^2)/3$
3. Repeat step 2 until the absolute difference between x and $g{\cdot}g{\cdot}g$ is sufficiently small.

Of course, you are not allowed to use the `Math.pow()` method for any part of this calculation, but you could use it to check the results of your work. Implement this algorithm first as an iterative calculation and then, in a separate function, as a recursive calculation.

6.12. Document 5.5 showed how to create two-dimensional arrays that can be accessed with row and column indices, rather than assigning "field names" to one of the dimensions. The code in that example showed how to populate such an array and then display it row-by-row. Add functions to this code that test the sums of each row, column, and main diagonals to verify that the specified assignment of integers 1–9 forms a "magic square." Then expand the code to create a 4 × 4 matrix. Arrange the integers 1–16 in this matrix so they form a magic square. Use the same functions to check your results. (This means that the size of the matrix must not be "hard coded" into the functions.)

6.13 A recursive algorithm for generating Fibonacci numbers is given in Section 6.5. Here is a variation that defines the totally obscure and completely useless "Brooks function" for positive values of n:

$B_n = 1$, $n = 1$ or 2
$B_n = 3$, $n = 3$
$B_n = (0.5B_{n-1} + 0.75B_{n-2})/B_{n-3}$

Give results for at least $n = 1, 2, 3, 4, 5$, and 20. You *must* use a recursive function to calculate values of the Brooks function.

| Give n: 2 | B: 1 |

Note: Running this script for large values of n (~100?) may cause your browser to lock up.

Extra Credit: Invent a new recursively defined function that is actually good for something.

Index

Printed in the United States of America